Mic Drop

James E. West as told to
Ilene J. Busch-Vishniac

Mic Drop

Autobiography of a Black Sound Scientist

With a Foreword by Gary S. May

James E. West
Johns Hopkins University
Baltimore, MD, USA

Ilene J. Busch-Vishniac
Baltimore, MD, USA

ISBN 978-3-032-20921-4 ISBN 978-3-032-20922-1 (eBook)
https://doi.org/10.1007/978-3-032-20922-1

Jointly published with ASA Press

This Springer imprint is published by the registered company Springer Nature Switzerland AG
The registered company address is: Gewerbestrasse 11, 6330 Cham, Switzerland

The ASA Press

ASA Press, which represents a collaboration between the Acoustical Society of America and Springer Nature, is dedicated to encouraging the publication of important new books as well as the distribution of classic titles in acoustics. These titles, published under a dual ASA Press/Springer imprint, are intended to reflect the full range of research in acoustics. ASA Press titles can include all types of books that Springer publishes, and may appear in any appropriate Springer book series.

The Acoustical Society of America

On 27 December 1928 a group of scientists and engineers met at Bell Telephone Laboratories in New York City to discuss organizing a society dedicated to the field of acoustics. Plans developed rapidly, and the Acoustical Society of America (ASA) held its first meeting on 10–11 May 1929 with a charter membership of about 450. Today, ASA has a worldwide membership of about 7000.

The scope of this new society incorporated a broad range of technical areas that continues to be reflected in ASA's present-day endeavors. Today, ASA serves the interests of its members and the acoustics community in all branches of acoustics, both theoretical and applied. To achieve this goal, ASA has established Technical Committees charged with keeping abreast of the developments and needs of membership in specialized fields, as well as identifying new ones as they develop.

The Technical Committees include acoustical oceanography, animal bioacoustics, architectural acoustics, biomedical acoustics, engineering acoustics, musical acoustics, noise, physical acoustics, psychological and physiological acoustics, signal processing in acoustics, speech communication, structural acoustics and vibration, and underwater acoustics. This diversity is one of the Society's unique and strongest assets since it so strongly fosters and encourages cross-disciplinary learning, collaboration, and interactions.

ASA publications and meetings incorporate the diversity of these Technical Committees. In particular, publications play a major role in the Society. The Journal of the Acoustical Society of America (JASA) includes contributed papers and patent reviews. JASA Express Letters (JASA-EL) and Proceedings of Meetings on Acoustics (POMA) are online, open-access publications, offering rapid publication. Acoustics Today, published quarterly, is a popular open-access magazine. Other key features of ASA's publishing program include books, reprints of classic acoustics texts, and videos. ASA's biannual meetings offer opportunities for attendees to share information, with strong support throughout the career continuum, from students to retirees. Meetings incorporate many opportunities for professional and social interactions, and attendees find the personal contacts a rewarding experience. These experiences result in building a robust network of fellow scientists and engineers, many of whom become lifelong friends and colleagues.

From the Society's inception, members recognized the importance of developing acoustical standards with a focus on terminology, measurement procedures, and criteria for determining the effects of noise and vibration. The ASA Standards Program serves as the Secretariat for four American National Standards Institute Committees and provides administrative support for several international standards committees.

Throughout its history to present day, ASA's strength resides in attracting the interest and commitment of scholars devoted to promoting the knowledge and practical applications of acoustics. The unselfish activity of these individuals in the development of the Society is largely responsible for ASA's growth and present stature.

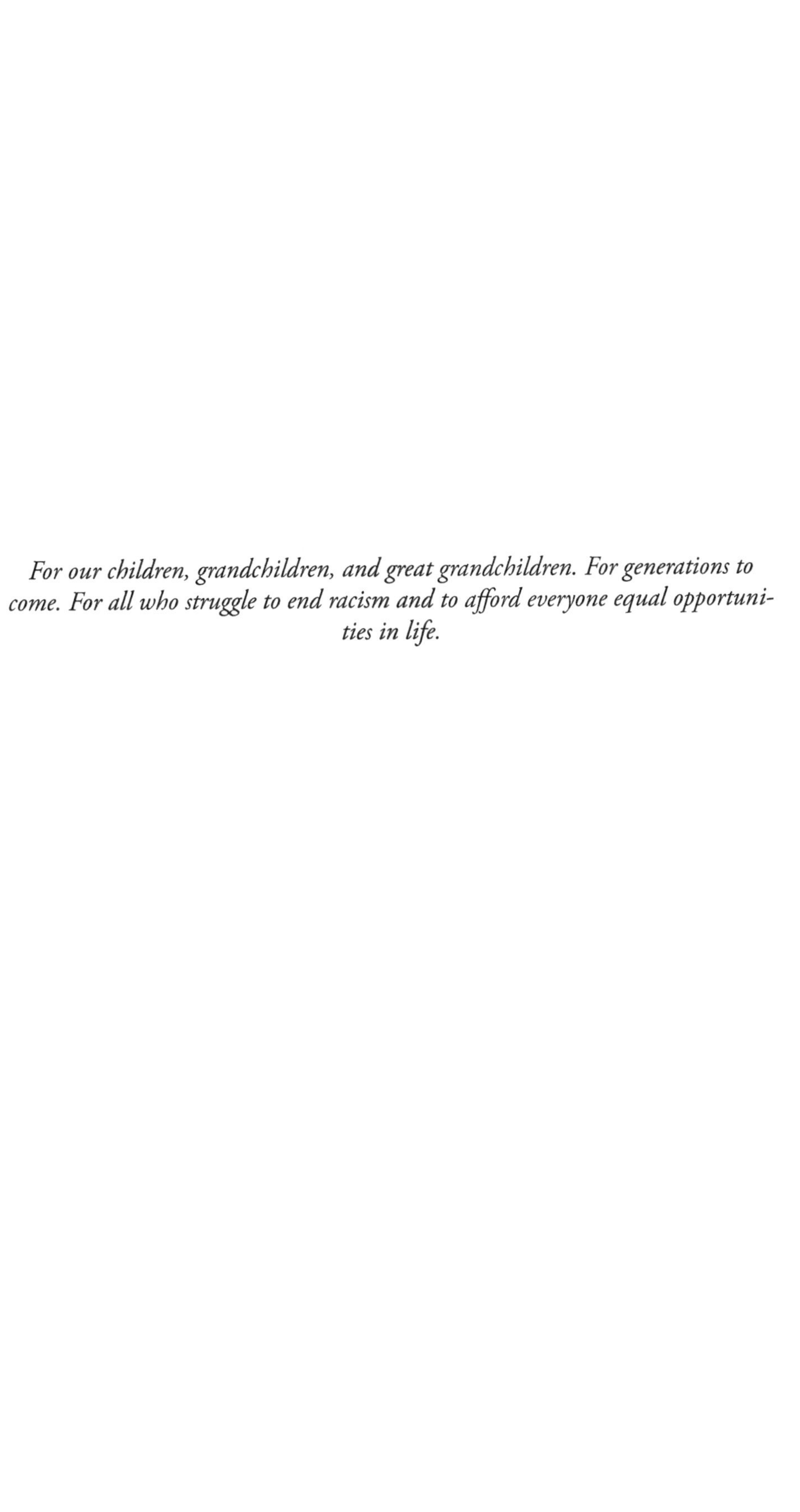

For our children, grandchildren, and great grandchildren. For generations to come. For all who struggle to end racism and to afford everyone equal opportunities in life.

Foreword

I've known Dr. Jim West for decades. He has been a hero and role model to me and to generations of others, not only because of his bold and groundbreaking scientific career but also for his advocacy and mentorship to countless young scholars, scientists and emerging leaders—part of his extraordinary work to increase the participation of underrepresented groups in the STEM disciplines. When I met Dr. West, I was a graduate student at UC Berkeley, and it didn't take me long to understand the critical role he was playing in the field of engineering. He was a dynamic and prolific innovator in the lab, and he was also systematically breaking down the doors that had previously been closed to himself and other underrepresented scholars and scientists. He writes, "I can look back now and be very proud of having participated in programs that aim to encourage women and underrepresented minority students to move into STEM careers along the entire pipeline length, from kindergarten through high school and college…For me, it is about exposure to the whole area of engineering, chemistry, mathematics, and sciences that may interest these kids. Knowing that there is a path for them heightens their desire to continue in the field."

As a student in the 1970s and 1980s, I had been one of those kids. My parents, Warren and Gloria May, were of a generation that grew up having to sit at the back of the bus. They both went to all-Black public high schools, but in the 1950s, as civil rights efforts were beginning to break down the system of racial segregation, my mom was one of nine Black students to integrate the University of Missouri. I grew up hearing about her experiences with racism and discrimination, along with those of other family members and friends. I learned to recognize them when they inevitably came my way as well.

In 1981, I entered the Georgia Institute of Technology as an electrical engineering student. During my time there, I would walk into a classroom or laboratory and was often the only Black student in the room. After graduating from Georgia Tech, I went to California to work on my M.S. and Ph.D. degrees at UC Berkeley. Again, I found that I was often the only Black student in my classes, and I questioned the reasons for that and began thinking seriously about ways to change it.

I first encountered Dr. West when I received a fellowship from AT&T Bell Laboratories in Murray Hill, New Jersey, where he worked. I had been hired one summer to work as a device test specialist in the High Speed Analog IC Group, where I also developed software systems to perform automated extraction of critical performance parameters for MOS transistors. Dr. West co-founded the Association of Black Laboratory Employees (ABLE) during his time there, and he was instrumental in the creation and development of programs such as the Summer Research Program and the Corporate Research Fellowship Program (CRFP) to support and provide opportunity for graduate students and minority students. CRFP awarded the fellowship I had received.

While working at Bell Laboratories, I witnessed firsthand Dr. West's deep commitment to mentorship and advocacy. He demonstrated for me and many others what it meant to actively shape better pipelines, build inclusive programs, and open doors for future generations. His leadership and vision in these areas inspired me to trust my own instincts to lead and to push for change. His example fueled my own commitment to support and mentor students throughout my career, especially those traditionally underrepresented in science, technology, and engineering fields.

Over the following years, after joining the faculty at the Georgia Tech College of Engineering, I created the Summer Undergraduate Research in Engineering/Science (SURE) program with a grant from the National Science Foundation. I also co-founded and directed the Facilitating Academic Careers in Engineering and Sciences (FACES) program to increase the number of African American Ph.D. degree graduates produced by Georgia Tech. Over 15 years, the FACES program produced 433 minority PhD's in science and engineering, which was the most in the USA over that time period.

Eventually, I became Dean of the College of Engineering at Georgia Tech in 2011 and then Chancellor of the University of California at Davis in 2017. At UC Davis, we have programs such as the Center for the Advancement of Multicultural Perspectives on Science (CAMPOS), which focuses on expanding the ranks of women and underrepresented faculty. Just as Dr. West did for me and so many others, these CAMPOS scholars now serve as role models and mentors to members of our campus community. Our College of

Engineering is also creating an innovation ecosystem, with numerous design clinics and programs to encourage student innovation and support multidisciplinary teams working on real world projects.

One of my favorite sayings is, "We rise by lifting others." And in this case, it couldn't be more accurate. None of the accomplishments I've listed would have been possible without the courage and hard work of Dr. West and the other visionary individuals who fought for change and cleared the way for us to reimagine the future. I built my dreams and goals on the foundation they created, and my responsibility is to ensure the next generation will have the opportunity to take things even further.

Dr. West mentions that when he has received recognition as the "first Black man" to attend or achieve something, it simply made him work harder to make sure more underrepresented people would receive that recognition in the future. His legacy, rooted in groundbreaking innovation and transformative social change, has had an extraordinary impact and continues to inspire and guide generations of scientists and engineers, including me. His story is not only a testament to his own brilliance and perseverance, but a call to action for all of us who follow in his footsteps, charting a course into the future while keeping our eyes on the unbound and limitless horizon.

Chancellor, The University of California at Davis
Davis, CA, USA

Gary S. May

Preface

Over the course of my career, and with increasing frequency as I've gotten older, I've had many people suggest that I should write my autobiography. Usually when this is suggested, it follows a story I've told which a friend or colleague believes marks me as extraordinary. My life story, these well-meaning people reckon, would serve as an inspiration.

I've resisted writing my autobiography in large part because I *don't* think of myself as extraordinary. I am instead an ordinary man who chose to pursue his passion for science and who overcame racist assumptions to succeed at acoustics research. I don't want to be seen as extraordinary. That there are few Black scientists who have been as successful as I am is rather a testament to the disparity of opportunities that have been afforded to people.

I write this autobiography to make it clear that a career in science or engineering is a viable alternative for people of all types. You don't need to be extraordinary to be successful in these fields, and success not only means being paid well for your work but also being able to enjoy it. It doesn't matter whether you are male or female, Black or white, religious or agnostic. Science doesn't discriminate except for right answers from wrong answers.

In my work I have always thought about how to make the world better in some way. Can we make communication by telephone clearer? Can we build a less expensive but higher quality microphone than currently available? Can we determine whether someone has pneumonia just from their lung sounds? For humankind to continue indefinitely, we need more people thinking about how to make the world better through technology. Importantly, I have learned that the best successes come from team efforts and from collaborations that involve people from different backgrounds. For that reason, diversity is our friend and not an enemy.

I hope this autobiography will inspire people of all sorts to consider a career in science or engineering. If it also helps people understand the history of racism and how it impacted lives, that is an added benefit.

Baltimore, MD, USA James E. West

Coauthor's Preface

I first met Jim West when I went to Bell Labs as a postdoctoral fellow in the Acoustics Research Department. I was told that I could work with anyone in the department, and I should take my time to identify the best project and collaborator for me. After talking with all the researchers, it was clear that Jim was the right match for me, although he didn't seem terribly excited at the idea of us working together. That certainly changed as we defined and started work on a project.

I have now worked with Jim off and on for nearly 45 years. We have published many papers together and been granted patents as a team. We've traveled together to all sorts of places to give talks, but research is only part of what defines our relationship. Over the years I have come to know Jim's wife, his children, his brother, and several nieces, nephews, and grandchildren. My husband and I have joined Jim and his family for Thanksgiving meals, and we have shared a variety of special occasions. We have ceased to be just colleagues and are now family.

I have been blessed with the opportunity to hold interesting jobs in a variety of places. I've been an academic administrator in Maryland, Ontario, and Saskatchewan, a faculty member in Texas, Massachusetts, Maryland, Ontario, and Saskatchewan, and an officer of various national professional societies in the USA and Canada. These positions have afforded me the opportunity to meet thousands of students and scholars along with a few Nobel Prize winners, authors, politicians, and performers. Among these many contacts, none has held my heart and my brain more than Jim West.

Jim's story is one of perseverance and triumph over a system that was (and still is) unapologetically racist. At every step, from early education in a rundown school, to the army, college, and employment, racism put obstacles in

Jim's path to success, yet he succeeded wildly and with grace. Despite having an excuse to be bitter about his treatment, Jim has maintained a pleasant demeanor and has radiated warmth. He is at his best when speaking with young Black students about the possibilities of science and engineering careers.

I wanted Jim's story of success in the face of racism to be available to inspire others. I knew Jim well enough to know that he wouldn't write his autobiography on his own, though, as he had too many other important things to do. Tooting his own horn never has appealed to Jim. For that reason, I pressed Jim to work with me to write his autobiography, and he agreed. We set up a schedule of regularly chatting about anything and everything, and I set about drafting his life story from the interviews.

Although I have worked with Jim most of my adult life, I learned things about him that I hadn't known before this project started, and Jim has been wonderful about letting me ask questions about personal aspects that are painful as well as those that bring joy. As I have seen the whole picture of Jim's life emerge, I have gained even greater respect for him than I had before we started this autobiography.

Writing this autobiography with Jim has been one of the most pleasant projects I've ever taken on. I am pleased with the result and hope that the audience for the book will similarly find Jim to be an interesting, inspiring, very down-to-earth scientist, father, colleague, and friend.

Baltimore, MD, USA Ilene J. Busch-Vishniac

Acknowledgments

This autobiography came together with the aid of many members of the West family, including my wife, Marlene, and my children, Melanie, Laurie, Jay, and Ellington, all of whom made time to tell stories about life with me. I am grateful for their willingness to share their father and husband with a larger community.

Through the course of my life, I have been blessed with wonderful colleagues at work. An incomplete list would include Ed David, Manfred Schroeder, David Berkley, Jim Flanagan, Gerhard Sessler, Gary Elko, and Ilene J. Busch-Vishniac. I am also very grateful to the Acoustical Society of America, the only professional society that has welcomed me and treated me as an equal to all other members.

This autobiography would not have been written without Ilene J. Busch-Vishniac taking the initiative on the writing. I am grateful that she saw the value of this project.

Thanks as well to Jill Tietjen for editing the book and making many great suggestions about content and format.

James E. West

Competing Interests The authors have no competing interests to declare that are relevant to the content of this manuscript.

Contents

1

Growing Up Curious

I've been driven by curiosity my entire life. Why do ants march in a straight line? How does a watch work? Why are Black children sent to a school far away when there is one close by? Curiosity has often led to my best discoveries, but it has sometimes gotten me into trouble as well.

I was born James Edward Maceo West in the winter of 1931, the second born from the union of my parents, Samuel Edward West and Matilda Omega Miller West. Their firstborn, a daughter, did not survive into childhood. I was born at the home of my mother's mother, Anne Miller, in Farmville, Virginia, where my parents were living at the time. That I was a home birth is not surprising, since the local hospital would not admit Black people.

My grandparents's home was built by my grandfather, James Peyton Miller, a skilled carpenter, and his sister's husband, Johnny Allen. The house was large, with four bedrooms, a living room (for Sunday use only), a gathering room, a kitchen, and a screened porch. Unlike many of the homes in the area, it had electricity and indoor plumbing. Grandfather and Great Uncle Johnny built a second home next door for Johnny's family. When I was born, that home also housed my cousin Edward Allen and his wife Vera. Eventually, our house was lost when the State of Virginia took the property in the 1950s for the Longwood College expansion and demolished the structure. I had wanted to fight that process, but I still had relatives living in the area at the time, and years of racism had taught them that nothing good ever came of bucking authority. My family also knew that if they fought, they wouldn't be able to continue to live comfortably in the community.

Because my maternal grandfather died before I was born, I knew him only through his portrait, which hung in the living room. I would look at the

J. E. West, I. J. Busch-Vishniac, *Mic Drop*, https://doi.org/10.1007/978-3-032-20922-1_1

picture and hold an internal conversation with him. Grandmother told me that she was sure I had a special connection with him, because I said true things about him that she had never mentioned to me.

Grandmother was the family matriarch and a mother of the church. She was very active in all church activities and was looked upon as a source of advice by the congregants and the church officiants. It was my first clue that Grandmother was a great person and worthy of respect.

Grandmother was largely responsible for raising me when I was very young. She was deeply religious and quite strict. Holidays were spent preparing food for those less fortunate than we. When not tending the children in the neighborhood, she would stay busy tending her garden and raising chickens. Her gardening and animal husbandry skills must have been learned rather late in life as she was born in Brooklyn, New York where there would not have been space for a garden or chickens. Her family migrated to Virginia, bucking the trend of Black families moving north. I know now that she worked cleaning the library at Longwood College, but as a child I thought she was the librarian. Thanks to Grandmother and the house Grandfather built, we lived comfortably. Our needs were always met, although luxuries beyond needs were rare. Grandmother was very giving. Nobody with a true need left her presence without some help.

Farmville, Virginia, is a small town that serves as the county seat of Prince Edward County, about 50 miles southwest of Richmond. Farmville was developed near the headwaters of the Appomattox River, largely to support the vital river transportation route. It had a large lumber mill. In the nineteenth century, a railroad connection was added to Farmville, but it is now a hiking and biking rail trail and a park. It was, at the time, a predominantly white community. When the Supreme Court ruled segregated schooling unconstitutional in 1954 in *Brown v. Board of Education*, Prince Edward County's government closed all its schools rather than be forced to teach an integrated student body. Then they opened private academies for white students and nothing for Black students. It wasn't until 1959, when the Supreme Court again interceded to rule the actions of Prince Edward County unconstitutional that public schools were reopened in Farmville. Farmville is best known today as the home of Longwood University. It is also the town closest to Hampden-Sydney College. Farmville is still small, with a population of a little under 7500 according to the 2020 census.

My mother was born in Farmville around 1900 and grew up in the house where I was born. She attended school in Farmville and sang in the church choir along with her brother, who became a deacon of the church. Mother attended, through a program of the Commonwealth of Virginia, Hampton Normal and Agricultural Institute (later Hampton Institute and today Hampton University), a coeducational and nondenominational school focused on training African Americans to be teachers.

Hampton Normal and Agricultural Institute was founded in 1868 by leaders of the American Missionary Association to educate freed Black men, although women were admitted as students from the early days. Its first principal was Samuel C. Armstrong, a man who had served as a Brigadier General for the Union. He modeled the school on the work of his father, who taught reading, writing, and arithmetic to Polynesians. Armstrong's stated goal was to enable Black men to become self-sufficient in the South. His approach mingled cultural affirmation with moral and manual education. He described it as education that involved "the head, the heart, and the hands." The school Armstrong founded stayed true to its mission and over the years expanded greatly.

At the time Mother went to Hampton Normal and Agricultural Institute, the only professional jobs open to women of color were teacher and nurse. Mother and Father were already married when she attended Hampton, and I was born before she finished her degree. To repay her state loan, Mother was required to teach at an Indian reservation near Wilkes-Barre, Pennsylvania for a couple of years. She considered the reservation to be so poor, dangerous, and depressing that she did not take me with her, instead leaving my grandmother in charge of my upbringing, as Father was frequently traveling. I recall that I missed her greatly and felt abandoned by her as she left to teach. Upon her return, Mother taught locally.

Father was from Green Bay, Virginia, about 15 miles from Farmville, where he grew up on the family farm. Like my mother's family, Father's family bucked the immigration trend and had moved south to gain land and begin farming. My paternal grandpa, Samuel West, was a minister who worked in three churches, and his wife was Fannie Mae West. Their home was more rustic than ours, lacking electricity and indoor plumbing. To distinguish my father's parents from my mother's, I called them Grandma and Grandpa. Mother's parents were Grandmother and Grandfather.

I describe my father as a "jack of all trades and master of a few." At different times in my life, I remember he worked as a funeral home operator, an insurance salesman, and, like many Black men of his time, a Pullman porter on the B&O Railroad. The railroad job kept my father away from home for weeks at a time. My father was older than my mother by a few years and he liked his drink far too much. I've occasionally explained the situation by saying he would awaken and brush his teeth with Seagram's 7.

My father's first wife died in an automobile accident. When I came into this world, I had two half-brothers: William (Sonny) and Leroy. They were 10 and 12 years older than me and were fond of teasing me by saying, "I changed your diapers. Don't you get fresh with me." Sonny and Leroy were largely raised by my paternal grandparents after my father's first wife died, although they also spent time with us in Farmville.

Unlike my maternal family, which was very dark skinned, my paternal family was quite light skinned. In Virginia at that time, you were considered Black if just 1% of your blood was from a Black person, but within the Black community it mattered how dark you were. This form of racism is called colorism, and it affected me greatly throughout my life. It has been suggested by relatives that the West surname was originally longer and came from a white, Jewish patriarch. Certainly, Father's skin was light, and his hair was straight rather than being tightly coiled as is common among Black folk. Similarly, Father's mother was light skinned and was either part white or part Indian. For purely personal reasons, I prefer to believe Grandma was part Indian. Unsurprisingly then, I came along with light skin and was dubbed a mulatto. The colorism of the time meant that growing up, the very dark-skinned neighbors shunned me and, of course, the white children didn't play with me. I tended to spend time with the other mixed-race kids.

Four years after I was born, my brother Nathaniel came along. Like many only children, I was none too happy suddenly to be required to share attention with this baby. And like many younger children, Nate suffered most from my irritation with the situation. Once, I remember I tried (unsuccessfully) to push his stroller down the steps while he was in it. I don't recall the particulars of what punishment I received, but I didn't try to get rid of Nate that way again. Of course, now we are grown and both in our twilight years. We acknowledge our love for each other, even if we spent many years trying to compete in various ways.

A couple years after Nate was born, cousins and next-door neighbors Edward and Vera Allen celebrated the birth of their daughter, Edwilda. Their second daughter, Edna, who I knew as Baby Sister, came along a year later. Cousins Edwilda and Baby Sister are shown in Fig. 1.1 in 2006. The four of us kids spent much time together and thought of ourselves as the Four Musketeers—out for adventure. Of course, as the eldest, I was generally held responsible whenever an adventure turned into a misadventure. Although I resented this at the time, I admit that as the oldest of the four of us, I tended to encourage and model risky behavior. For instance, my maternal grandfather's shop was one of my favorite places to get lost. Although Grandfather had passed on many years before, his shop and some of the tools were kept in

Fig. 1.1 Edwilda (left) and Baby Sister (right) in 2006

place and I would often disobey the rule not to enter the space so that I could play with the tools that remained.

Grandmother set an example of reading for us. She read the Bible and *The Farmville Herald*. I don't remember how old I was when I picked up the newspaper and pretended to read it. On that occasion Grandmother wouldn't read the paper aloud to me and I only found out why much later. While I couldn't read the text, I was perfectly able to understand the pictures and that day's newspaper showed a white woman in a white robe admiring the lynching of a Black man. Today, I am reminded of those photos by the images of the treatment of inmates at Abu Ghraib, the facility where prisoners captured early in the Iraq war were tortured by their US captors.

Mother also read to me, and I have vivid memories of sitting with her by the coal stove in the house. Sometimes Mother would talk with me about what could be rather than what was. It served as my inspiration for much of my life.

When I was old enough to go to school, I went to the local elementary school set aside for Black residents. My school was old and poorly resourced. The bathrooms afforded no privacy—an issue Edwilda and Baby Sister dealt with by training their bodies, so they never had to use those bathrooms. My route to the school took me past the elementary school for the white residents of Farmville. It was new and shiny and filled with all sorts of resources, including new textbooks. By contrast, the school for Black students used the old textbooks cast off by the school for the white students. As a child, I wondered why we were sent further away to an inferior school. I didn't fully understand racism and its most obvious symptoms—segregation and unequal treatment.

Because Mother was a teacher locally, as was my cousin Vera, I found that my mother knew almost all my teachers. As a child, I was convinced they were all relatives of mine, because they treated me with the familiarity of a relative. I now understand that even those who weren't relatives (and many were) were under orders from Mother to keep me on the straight and narrow. The neighborhood communication was sufficiently good that, as Nate has reminded me, on days when I knew I had done something that I shouldn't have done, I would stop along the way home to pick up the switches that would impart punishment, so that I could get that part over with quickly.

My childhood doesn't strike me as unusual in any way. I spent lots of time outdoors, letting my curiosity drive my interests. Mostly, these were benign activities. For instance, I recall watching a line of ants working hard to carry food to their nest and wondering what would happen if obstacles were put in their way. I put some twigs across their path and watched them move over them and keep going, leaving me impressed by their single-minded persistence. I climbed trees and examined anything I could find for clues of how the world worked. I was fond of taking Nate's toys apart to see how they functioned. Often, I couldn't reassemble the toys.

One of the striking characteristics of growing up in Farmville is that we children understood that we were free to wander. We knew to stay within the Black community and to watch out for cars as we walked, ran, or bicycled along the roads. We knew to return home in time for meals and chores. By the time I was 8 or 9 years old, I would cruise the neighborhood on my bicycle looking for new things to discover. If I saw a telephone repair truck, I knew that Mr. Styles, the local repairman, was somewhere close and I would look for him. Mr. Styles would regularly patch the holes chewed in the lead sheathing of the telephone wires by squirrels, and I would fire questions at him like bullets from a gun. He would patiently answer them and encourage me to think of more questions for the next time I saw him. Mr. Styles was an inspiration. He was the first technologically minded person I met, and he encouraged me to learn.

Living near the Appomattox River, I learned how to swim the old-fashioned way. My father threw me into the river and told me to swim. While this might seem cruel by today's standards, it worked for me. I took to the water quickly and became a great, self-taught swimmer. Swimming in the river meant I had to learn how to deal with currents, and I became strong and fast. To this day, I prefer to live near bodies of water, as they have always felt like home.

From a relatively early age, I was fond of taking things apart to figure out how they worked. This sometimes got me into trouble. If a pair of pliers or screwdrivers could be found and there was something that I could open, it got opened. And it did not always get back together in a working condition. The most difficult situation for me was when I took my grandfather's pocket watch apart and could not get it back together again. It had more than 100 pieces. That was rather unfortunate because it was a nice Hamilton watch that I destroyed. I was severely punished for ruining this watch.

My family knew that it was important to keep me busy and monitored so I avoided trouble. After school, I went to my Uncle Nathan's dentistry office, where I would help him clean up. Nathan was my mother's brother. My uncle mixed his own fillings, and the materials used at that time included mercury. Inevitably, some would land on the floor in the mixing process. I loved playing with the mercury since it behaved uniquely, but I especially loved how it could be used to coat things. I started a small business of coating pennies with mercury and selling them to people as dimes. That was the start, and nearly the end, of my young entrepreneurial career. My parents made me fold my operation and apologize to those I had scammed.

Uncle Nathan bought his dental supplies through Gray's Drug Store in Farmville. I remember that there was a soda fountain there and sometimes I would go with Uncle Nathan when he went to pick up supplies. I would be treated to a chocolate malted milkshake, the absolutely best taste I've had in my life. One day, Uncle Nathan asked me to do some business for him at the post office, so I did and then went to the drug store. I hopped up on the stool and told the person behind the counter that I wanted a milkshake. Service was refused by the counter worker. That's the first time I ever got called a nigger ... by a white person, anyway. I stood up and said, "is Dr. Gray here?" expecting him to correct his staff person. Dr. Gray looked up from behind his counter and he didn't move a muscle. After that, my uncle and my aunt drove to Richmond, about 50 miles away, for their supplies.

If time permitted after helping Uncle Nathan clean his office, I would visit Dr. Baker, the local general practice doctor next door to the dentistry office. Dr. Baker made house calls, and I would sometimes travel with him to see a patient. Dr. Baker was more like Jesus than anyone I knew. He took care of the poor and was extremely gentle and kind. When patients couldn't pay in cash, he worked out something as a barter transaction. I considered him a mentor.

When I was about 11 or 12 years old, Dr. Baker died, and nobody could explain to my satisfaction why he was gone. Up until that point, I had been a regular churchgoer with my family, but I stopped going to church when Dr. Baker died. My family tended to use punishments rather than rewards to impact behaviors, so my insistence that I would not go to church prompted punishments. My grandmother, in particular, was very concerned about me ceasing to go to church. They even had the head of the church come to speak with me, trying to explain death and God's will. Nonetheless, I remained steadfast in turning my back on religion. I have never returned to church-going ways, and I have never regretted that decision.

The only exception to my heathen ways would occur in the summers when I would spend time at my paternal grandfather's and grandmother's farm. There was simply no way that a minister would let the children in his house avoid Sunday church. Most Sundays we went to two of the three churches where Grandpa was a minister. The churches were small rural congregations. Parishioners arrived on foot or by horse and buggy. Typically, there would be one or two cars in the parking lot, and one of them belonged to Grandpa. Through his preaching, it became clear that Grandpa was not a fan of Franklin Roosevelt. He simply didn't believe Roosevelt was a friend to Black people. Grandpa was a faithful Republican, committed to the party of Lincoln.

A fair number of crops were grown on Grandpa's and Grandma's farm. Almost everything we needed to eat was grown on the farm, much of it stored over the winter in Mason jars. Tobacco was the cash crop, sold for income. As Grandpa and Grandma had the only barn in the area, all the local tobacco farmers would cure their tobacco in Grandpa's and Grandma's barn. Curing tobacco meant that the firepits had to be tended 24 h a day. Neighbors would come over to tend the fires with their harmonicas and guitars, and there would be nonstop music and dancing, which I very much enjoyed. I expect Grandma and Grandpa charged hardly anything for this, seeing it as a service to the community.

Part of our routine at the farm was to go into the town center on Saturdays. In the summer, when we went into town, we would get the corn that had been harvested ground at the mill. A few items that we couldn't grow, mostly for the animals on the farm, would be purchased. I believe, though, that the true purpose of these Saturday trips was to be sociable. Town was a gathering place for the residents of the area, and it gave us a chance to see our neighbors. It was certainly how I met the other children in Green Bay.

My first summer at the farm, my parents dropped me off and then went home. Mom came back a few weeks later to see how I was getting along and ended up in tears. My beautiful curls had all been shaved off to prevent lice, a chronic problem on the farm. I was clearly less worried about this than Mother was.

Summers at Grandpa's farm were spent with Leroy and Sonny as well as the children of my father's brother, Alexander: Alexander, Jr. and Mataw West. I became close to all of them. There was a creek running behind the farm that had been dammed to create a swimming hole. Between this and the woods and the animals and the garden, we had lots of room to play. The six of us would manage to get into all sorts of trouble after doing our chores. One example of trouble I caused related to the well feeding water to the house. I decided to test how deep it was using something like a two-by-four wood plank, but the wood slipped from my hand and fell into the well and blocked the water outlet. Ultimately neighbors came over with chains and after about 3–4 hours of work they managed to clear the blockage. Grandpa would write our misdeeds in a notebook, presumably to determine what punishment was called for over time, but I never suffered punishment for the well pump disaster or any other of the lapses Grandpa noted.

When I was old enough, Cousin Edward Allen used me in his business of electrifying homes. The process of bringing electricity to homes required stringing wire from end to end of the house. I was sent to crawl under the homes to string the wire. It was my very first experience with electricity and set me on a course to become an electrical engineer. I learned what was safe and what was not, although I didn't always use the highest of safety standards in my work. For example, I became proficient at taking apart electrical sockets without turning off the electrical power.

I will admit that not all my experiments and experiences with electricity were benign. My brother, cousins, neighbors, and I would all scour the properties around us for anything useful being tossed out. Once I found a radio someone had thrown away. It was a nice, steamy August day in southern Virginia, and the only outlet in my bedroom was in the ceiling. I stood on my brass bed and plugged in the radio only to feel the electricity coursing through me. I got stuck there until Leroy knocked me off the bed. I was apparently none the worse for wear, but I like to think I became a bit smarter about what limits I should set on electrical experiments. It did convince me that I needed to better understand electricity.

Working with Cousin Edward also introduced me to the realities of racism. Edward was honest in a way other family members weren't about racism and what to expect. Father had told me not to expect the police to be friendly, and to cross the street if I saw a policeman on the side I was walking on, but that was the limit of what he would say. I later came to understand that nearly every Black man I grew up with had a police record and regularly got stopped for doing nothing in particular. Edward told me more about the roots of racism and how it would manifest throughout my life. Through Edward I came to understand that our community viewed Black men with distrust and strove to limit the jobs available to them. To survive and to support their families, the Black men in my family had learned to do the foot shuffle, the motion indicative of subservience.

My father made it clear that this foot shuffle was what he did to survive and not what he wanted to do. When we would encounter a white person, he would immediately resort to the shuffle and respond to all questions with a "Yes, sir" or "No, sir." He wouldn't flinch when he was called a "Sammy" or I was named a "pickaninny." But when that person was out of earshot, Father would curse for at least 3 minutes. It's how I learned to curse.

I've never been much of a reader, so I was not likely to be found outdoors under a tree or by a brook with a good book in hand. Back then, they didn't diagnose and treat dyslexia, a reading disorder I now recognize I have. I can still remember the fear of being called on to read something. I knew that this fear set me up for failure, that it was self-fulfilling, but I couldn't control it. I'd be fine for the first two or three words, then I'd start screwing it up. My dyslexia is sufficiently intense that I've never been able to effectively use written notes for a presentation I am to give.

Despite my reading problems, I recall with great affection a book an aunt from Buffalo gave me about Benjamin Franklin when I was quite young. I wanted to try his experiment with a kite in a thunderstorm and to be like him in pursuing answers to the questions the world presented to me. Many years later, when I was awarded the Benjamin Franklin Medal by the Franklin Institute, I felt that my life had come full circle, from reading about Benjamin Franklin to being honored for being like him.

When I was 12 or 13 years old, the family moved to Hampton, Virginia, where the schools open to us were much better. Had we not moved, I would have attended the Robert Russa Moton High School in Farmville. That school had no gym, no cafeteria, no blackboards, and no indoor bathrooms. There were no science labs, and the general lack of space meant some classes were taught in a school bus.

Nate and I were able to finish our grade school at the George P. Phenix High School, a school established for Black students at Hampton Institute. The school was founded in 1931 and named for the fourth principal of Hampton Institute, the person who lobbied the Commonwealth of Virginia to allocate funds to build a modern school for the youths of the city. Unfortunately, George Phenix drowned a few months before the school opened.

The Phenix Training School and High School were intended to serve as a teaching laboratory for students at Hampton Institute, who gained experience as student teachers at the school. The Phenix School was very well equipped, including shop classes, home economics classes, and a business area with typewriters. The school had the only symphonic orchestra among the high schools for Black children in Virginia. It was an entirely different world from the schools I had attended earlier, and I thrived in the new environment. I spent time helping the chemistry and physics teachers after school. A friend and I also built our own telephone system, or perhaps I should say mostly built, as we were caught before it was finished.

The telephone incident started with toy phones that actually worked. They needed to be wired, but once that was done my friend Paul Robinson and I could talk between rooms in the house or between the garage and the indoors. We loved this ability to speak on our phones because using the real telephones was forbidden by our parents. Paul and I looped in another friend, Calvin Thomas, and we decided we would connect our homes, so the toy phones worked for us even while we were at our respective houses. We knew there was a telephone warehouse not far from where we lived. We found a way to break in and steal some telephone wire—enough to connect our homes. We strung the wire using trees to support it where we could and running through the backyards of neighbors, and of course we used the telephone poles to hold the wire as well so we could cross streets. That was our downfall. The telephone company saw the lines and traced them back to my house. We hadn't yet completed the connection to Paul's or Calvin's house.

In the 1960s and 1970s, the phone company often forgave people it found trying to embellish or skirt around their system, even hiring the more enterprising of the people they caught, but this was not the case in the 1940s. The phone company threatened to put me in jail, but I was only 14 years old at the time, so I think that was just a threat to scare me. If so, it worked. I had been the recipient of multiple lectures about not trusting the police and now I was being threatened with jail. My father interceded and got the charges dropped but I was in big, big trouble for stealing and for invading other people's property without permission.

It was during high school that I came to understand that I learned much more efficiently on my own than by trying to understand what the teachers were telling me. In addition to dyslexia, I think I was probably suffering from ADHD. I had an abundance of energy but a limited attention span. It made formal class settings painful, and I ultimately avoided them as fully as possible. But there was also something special about my way of learning. I remember, for instance, that my father could hear a tune and then play it on the piano. I could do that as well, but I wouldn't just regurgitate it—I'd embellish it. I could do the same thing with science. I would read and understand something and come up with more than was presented to me.

My high school math teacher, Dr. Gladden, lived next door to my family. I always felt that he and my mom had some signals that I was not aware of because, once I finished my homework and chores, I was able to hit the back door to go play for a little while. Dr. Gladden was frequently right there waiting for me. He'd choose a little patch of soil where he would draw geometric figures and talk about geometry. I could do calculus before I knew what the word was, before I had any idea of what that was all about. I understood the principles because this was just drilled over and over and over in me by Dr. Gladden.

Once we were settled in Hampton, Mother found a job as one of the human "computers" at Langley Research Center next door to the Langley Air Force Base. These computers, featured in the book and movie *Hidden Figures*, worked on the mathematics required for space orbits. Mother worked specifically on satellite navigation systems and loved her job. She had followed Dorothy Vaughan, the oldest of the Hidden Figures in the book and movie, a distant relative who also moved from Farmville to Hampton, into the computer program.

At this time, World War II was raging, and Joe McCarthy had emerged as a loud voice against communism. McCarthy and his colleagues had gathered a long list of organizations that were suspicious. Membership in any of them was likely to make your life difficult, with a strong possibility of being fired from whatever job you had. Although the NAACP was not on the list of subversive organizations, McCarthy hated the organization and claimed its goals were contrary to the aims of America. Mother had become active in the leadership of the local NAACP chapter, focusing on employment for Black people, and pressure was brought to bear on the Langley bosses to remove her. She was fired in 1951 and never thereafter gained meaningful employment. Mother mostly worked as a nanny for the children of white families from that time forward.

One of the very interesting things about activism is that many activists didn't set out to create a ruckus or to markedly change a system. Such was the case for Mother. Having grown up in a family that believed in service to the community, Mother sought a means to continue that tradition. The NAACP at the time was an organization that focused on service to the Black community, so it was natural for Mother to become involved. Communism and its political ramifications had nothing to do with it.

But activism has been a common thread running through my family history. My cousin Vera Allen was involved in the Martha E. Forrester Council of Women, an organization that focused on harmony among the races and that fought for equal access to high quality education for Black and white students. Her elder daughter, Edwilda, was part of the 1951 student walkout at Robert Russa Moton High School in Farmville. In that walkout, 400 students marched to the courthouse to protest the conditions of the school, which stood in stark contrast to the conditions at the schools for white students in the district. Edwilda was one of the students who went into the courthouse and presented their grievances. The NAACP took note of this protest, which led to *Davis v. County School Board of Prince Edward County*. That legal case was the largest of the five combined to form *Brown v. Board of Education*, the famous case brought to the Supreme Court that found in 1954 that segregated schooling was unconstitutional. As a result of Edwilda's involvement in the walkout, Vera lost her state teaching license. She relocated to North Carolina where she worked for 5 years before returning to Virginia.

My own significant involvement in activism came later than Edwilda's and, likely thanks to the changes in our society, resulted in fewer penalties for my family and me. My first and not very successful foray into activism came while I was still at the Phenix School on the Hampton Institute campus. Marian Anderson, a famous Black opera singer, was coming to visit and perform at Ogden Hall. The event was aimed at donors, a group who were disproportionately white. Black students and faculty were not invited to attend. I used to sneak into Ogden Hall, and I knew how to avoid detection by climbing through a part of the ceiling into the performance hall. I did this regularly to avoid paying for shows I wanted to see. I gathered a couple of friends and suggested we disrupt the Marian Anderson performance by sneaking in and being raucous. Unfortunately, my plans were discovered, and I was severely reprimanded as my actions could result in fewer donations to the institute. To this day, I don't understand the logic of inviting a Black person to be on stage but then refusing to allow other Black people to attend the performance.

My early life was not atypical for a Black male of the time. I was free to explore within the confines of the Black community and quite aware of the difference in opportunities and treatments for white children compared to Black children.

2

College, Here I Come

Throughout my childhood, my family made it clear that they expected me to go beyond a high school education, to go to college, preferably to study to be a doctor or a dentist. My family understood that the employment options for Black men were limited, and that medical and dental careers reflected the best financial options for support of a family. Uncle Nathan, the Farmville dentist, hoped that Nate and I would both become dentists and join him in the practice he had opened.

Although it was unusual for Black students to attend institutions of higher learning at the time I was college age, education beyond high school was the norm in my family. My mother had a college degree as did cousins Vera and Edward Allen, Uncle Nathan, and my early teachers who were relatives. My father was the exception as he likely didn't even complete high school. I know that it is thanks to this focus on attaining education that my family was able to be financially comfortable even if we weren't what anyone would call affluent. Ultimately, Nate earned a dental degree. Baby Sister earned a PhD in instructional pedagogy, and Edwilda earned a degree from Alverno College. I left college before earning a bachelor's degree, but I have been awarded honorary PhD degrees in recognition of my research.

Earning college degrees was none too easy for Black people. Segregation tended to limit Black students to historically Black colleges and universities (HBCUs) which accounted for roughly 80% of the enrollment of Black students until the mid 1960s. HBCUs were not well funded at the time I was considering schools and have continued to be underfunded today. This historic underfunding means that the educational resources available at the HBCUs have been fewer than at the other public schools. It translates into an

J. E. West, I. J. Busch-Vishniac, *Mic Drop*, https://doi.org/10.1007/978-3-032-20922-1_2

enforced inferior educational experience for students at HBCUs compared to the other land-grant schools.

I applied for admission to a few of the in-state universities. Virginia Polytechnic Institute and State University (now commonly called Virginia Tech) was founded in 1872 as the white land-grant institution, with Hampton Institute being the slightly earlier founded Black land-grant university. I considered both Virginia Tech and Hampton as potential places to study. Virginia Tech made a point of letting me know that I couldn't possibly handle their curriculum. This made me doubt my own abilities, which is exactly what racism is supposed to do. It was not the first time that I found the expectations of people of color to be set very low and it wouldn't be the last time I confronted this bias. What I didn't know at the time was that Virginia Tech had not yet enrolled a Black student at the undergraduate level when I was looking at schools. It was in 1953 that Irving Linwood Peddrew, III became the first Black student to be enrolled as a freshman there.

Given the pressure from my extended family, I enrolled in Hampton Institute as a pre-medical student after graduating from high school. To keep the financial burden manageable, I signed up for ROTC (Reserve Officers' Training Corps), which provided a scholarship along with the commitment to military service. ROTC programs were very common at the HBCUs, providing a means of making college affordable and entrée into the officer roles of the military, which were otherwise not available to Black men.

Signing up for ROTC seemed natural in my family, which has had a history of military service. One of my mother's brothers, Uncle James, was killed while serving in France during World War I. I was named for him. Mother's brother, Nathan, also served in France during World War I. My first cousin, Walter Brown, flew with the Tuskegee Airmen in World War II. The Tuskegee Airmen were a primarily Black unit of military pilots and their support staff, famous for their extraordinary combat record while protecting American bombers from the enemy. They are also known for the unethical experiment to determine what happens with untreated syphilis. That experiment was carried out from 1932 until 1972, long after penicillin was established as standard treatment for the disease. I don't remember much about Cousin Walter, but I know everyone said he was a changed man when he returned from service. Other generations of the family continued this tradition of serving in the Army. My half-brother Leroy was drafted into the Army. My cousins Alexander and Mataw both served in the Navy. Ultimately, Nate volunteered to serve, and I have a grandson who joined the Army. I wasn't pressured to sign up for ROTC, but it didn't seem odd to my family.

At Hampton I swam competitively and enjoyed the camaraderie of the swim team. I had not had the benefit of swimming lessons before going to college, but I was so fast I could beat most of the swim team in races, so the team took me on and trained me. It was one of the most pleasurable aspects of my time at Hampton. But I realized quickly that I was not on a career track I loved. I couldn't understand why I was being forced to learn about all these plants and animals I had no interest in. I decided I needed to change schools and change my focus.

This unwillingness to waste my time learning topics I have no interest in has been a common thread throughout my life. In grade school I had a spotty record. If I found the topic interesting and I wanted to earn an A, I would earn the A. If I found the topic uninteresting, I would figure a C was adequate. I learn a lot with hands-on work, but I also do much better when presented with a book and a chance to learn on my own than I do with a traditional lecture style class. Partly the problem for me is my dyslexia, but also, I would get bored with the slow pace of learning in a traditional classroom. Given the right resources, I could fly through material when it struck me as interesting.

Nate, who worked hard to earn his good grades, thought I was cheating when I would get a good grade with minimal effort in a topic I enjoyed. We had different interests, but we were very competitive, and my easy successes bothered him. It was only when we were both much older that we realized we no longer had a reason to compete.

I discussed my desire to pursue a career in physics with my family but did not receive their support. My father introduced me to three Black men who had earned PhDs in physics and chemistry. None of them had been able to find jobs at universities because of their race. They were working instead at the post office, far removed from their aptitude and training. My parents said they would cut off financial support if I chose to get a degree that would not lead to a career in medicine or dentistry. Nonetheless, I stood firm in my desire to study physics.

I knew that Wilberforce University offered scholarships to its best swim team members and had an Army ROTC program, so I applied to transfer there and got accepted. Wilberforce University, a liberal arts university that was founded in 1856, is the oldest, private, historically Black university owned and operated by Black people. It was established at a time when it was illegal to educate Black persons, reflecting a bold decision by men and women of good will to transcend the limitations of racial policies and was named for the abolitionist William Wilberforce, who famously said, "We are too young to realize certain things are impossible…So, we will do them anyway." I determined that I would attend and pursue a degree in physics and continue my Army ROTC training. Alas, this was not to be.

When I left Hampton Institute for Wilberforce University, I transitioned from a school that worked on the semester system to one that operated on quarters. That meant I was going to have a gap before I could restart my education. Unbeknownst to me, this gap in my ROTC training meant that I was immediately eligible to be drafted, and I was. Had I known this, I would not have made the transfer at that moment, but nothing I tried managed to change my military service situation. I was now in the US Army.

3

You're in the Army Now

When I was drafted into the US Army in 1951 the United States was involved in the Korean War. Fighting had begun in 1950 when Soviet-supported North Korea invaded South Korea. The UN forces that were sent there were almost all US troops. It was into this foreign war that I was tossed.

I don't remember much about boot camp because, although I was a private, I was exempted from a good bit of what the recruits did, given my ROTC experience. I was given the job of teaching the troops how to jump from a platform (simulating a ship) into a pool while wearing a pack and life vest and making their way to shore. This was a fine activity for me, but it didn't last long. I shortly found myself headed to Korea to join troops on the front line. It was my first time traveling far away from home.

I arrived in Korea in the winter and learned it got very cold in Seoul. Seoul was an interesting city, but I didn't get to see much of it. I did learn that the local inhabitants we saw were willing to do anything for American dollars. For instance, one enterprising person was fond of taking whiskey bottles, drilling tiny holes in them to extract some of the whiskey, replacing the whiskey with water, blocking the holes with wax, and then selling it as if it were pure. I was sent north of the 38th Parallel, to Christmas Hill, a mere 5 days after landing in Korea.

Christmas Hill, also known as Hill Number 1090, was under fire nearly constantly from North Korean and Chinese troops for years. It was the site of one of the bloodiest battles of the Korean War, coming to a head just 2 weeks before the war ended. In that battle, more rounds of ammunition were fired toward Allied troops than in any other battle of the war as the Allies defended the hill against 80,000 enemy troops.

J. E. West, I. J. Busch-Vishniac, *Mic Drop*, https://doi.org/10.1007/978-3-032-20922-1_3

The US military services were segregated through World War II. President Truman formed a task force to look at the issue of segregation and discrimination in the Armed Services. After the task force produced its report, he issued an executive order in 1946 to end military discrimination. This order stated, "There shall be equality of treatment and opportunity for all persons in the armed services without regard to race, color, religion or national origin." Military leaders did not like this order and delayed making changes. The Air Force was the first service to begin integration, but it was a process only completed in 1952, 6 years after the executive order was issued. The Army's training units were integrated in 1951, and it was a fully integrated service in 1954.

Based on the policies of the military, you might think that I served my country in a Black unit, but this was not the case. I was the only non-white member of my unit. I think I understand how this happened. As a mulatto, I have always been a light-skinned Black person. Further, my skin tone has always lightened considerably in the winter months. When you get drafted, they also cut off almost all your hair, so the combination might have made the person assigning troops to units think I was a white man. As spring and summer arrived, it became clear that I was Black, but this didn't seem to impact my comrades greatly. Each of us had the same equipment, including a gun, and that seemed to equalize the relationships. In one skirmish, I helped one of the members of my unit, making sure he got back safely. He was grateful and told me later that I should come visit him in Biloxi, Mississippi, his hometown. I told him that couldn't happen, and he said he'd stand in main street and throw his arms around me and tell everyone how great I was. Nonetheless, I never did go see him at home. Don't get me wrong. There was certainly racism in the army, but it was different. We depended on one another, and I could give as good as I got without worrying about the reaction.

Christmas Hill was my first exposure to combat's reality. I saw a mine field with body parts still lying in it, and nobody willing to try to retrieve them for fear of setting off more mines. I learned to live with the constant noise of gun fire, and the anxiety of living in a war zone. I saw people wounded and killed. I killed a few enemy soldiers myself.

We had a series of elegant trenches and underground bunkers where we spent our time. We had nothing to keep us busy except checking on the enemy to be sure they weren't moving toward us. We mostly stayed in the trenches because anything and anyone that popped up drew fire. It created a nerve-wracking state of constant tension mixed with boredom.

Arriving in Korea, I was pretty clean. Given the experience of my father's drinking, I didn't regularly consume alcohol and certainly not to excess when I did. In the army, drugs and alcohol were the norm. They took the edge off the boredom and the anxiety. We never knew when we would be involved in a skirmish, and the danger was very real. It was during this time that I first used marijuana. My fondness for this drug lasted the rest of my life.

On Christmas Hill, the bunkers had facilities where we could cook meals. I decided that next time I had an opportunity, I'd go into the town and find a woman to do our cleaning and cooking and service us in other ways. On the way into town, as taught by the medic, I picked up about 10 doses of penicillin just to make sure that my fellow soldiers and I avoided communicable diseases. Like drugs and alcohol, sex was just another way to find comfort and decrease the boredom.

One day, while on patrol, one of the soldiers in my unit stepped on a Bouncing Betty, a grenade that pops out of the ground and detonates in air to cause as much damage as possible by spraying shrapnel far and wide. I took some shrapnel in my right upper hip. I was rescued by a medic and sent by helicopter back to Seoul and then on to Osaka for treatment and rehabilitation. This kept me away from combat for 3 months, but as soon as I was able to walk again, I wanted to go back to my unit. I understand this now as contradictory. I tried everything I could think of to avoid serving in combat, then when I was removed from it, I wanted desperately to return to it. I returned to my unit with a Purple Heart and a promotion to corporal.

Christmas Hill had a regiment of Canadian troops who spoke French. I had had 3 years of classes in French in school, so I was chosen to translate. Of course, my school French was the language of France and not the Québécois of Canada, so my translation ability was limited. I did my best as I was the only option available. For this assignment, I was given a jeep so I could move between the Americans and the Canadians. Driving was a part of the job that I enjoyed. I had developed a passion for fast driving and certainly speeding between posts was the order of the day, as I was particularly vulnerable to attack while on the roads.

On one of those trips between posts on exposed roads, the jeep was blown out from under me. I was once again injured and sent to Osaka to heal. Most of the damage this time was to my left hand. I lost the top joint of my index finger, and my thumb was somewhat mangled. This time, I took my second Purple Heart and headed for home.

The army was lifechanging for me. I learned a lot about myself while learning survival skills. I learned, for instance, that I work far better when left unfettered than I do when regimented and driven by a routine. I learned that I could live happily with far fewer material possessions than I had previously aimed to amass. I learned how to live in a community of white men without feeling a stark sense of imposed inferiority.

I returned home with the thought of once again pursuing a college degree in physics. I knew that my parents were still adamantly opposed to my choice of major, but I was equally fixed in my desire.

4

A College Man Again

Upon returning from the army, I was anxious to renew my college studies as soon as possible. Back then they didn't know much about PTSD (post-traumatic stress disorder) and the dangers of not coming to terms with trauma in your life. I just refused to think about my experiences and ended up with my digestive system going haywire, a problem that has plagued me since then. At one point, shortly after returning, I ended up with ileitis, an inflammation of the final segment of the small intestine, and I was hospitalized briefly.

After reviewing my options, I chose to attend Temple University with the GI Bill covering a significant amount of the cost. The GI Bill is formally the Servicemen's Readjustment Act of 1944 and mandated a variety of benefits to servicemen returning from World War II. These included low-cost mortgages, low-interest loans to establish a business or farm, a year of unemployment compensation, and payment of tuition and living expenses to attend high school, college, or a vocational school. The Veterans Readjustment Assistance Act enacted by Congress in 1952, ensured the benefits in the original GI Bill were extended to veterans returning from active service in the Korean War.

I had an aunt in Philadelphia who was willing to let me live with her and would feed me as long as I was productive. That put Temple University within financial reach, and I enrolled as a physics student. However, Temple University was not a perfect choice for a couple of reasons. Temple was best known as a liberal arts institution, so its science facilities were old and neglected. Further, not being an HBCU, Temple had a very small population of Black students—besides me there was one other Black physics student.

Temple University's culture revolved around the idea of study groups with students in a group taking classes together and studying together. Such cohorts

J. E. West, I. J. Busch-Vishniac, *Mic Drop*, https://doi.org/10.1007/978-3-032-20922-1_4

have been shown over the years to improve the performance of nearly all students in the group. But the study groups in my department rejected me because of my race. Taking a page from the women in my family, I decided to show the white students what they were missing by not including me in their group. I made a point of letting them know how I solved the problems we were assigned and demonstrated my ability to solve complex equations. Eventually that earned me invitations to join the very groups that had initially rejected me.

Living in Philadelphia, I saw a new perspective from the standpoint of poverty and the deplorable situations in which some people are forced to survive. I lost confidence in the system of justice and government in the US. I became a fan of Trotsky and believed that change would require the violent confronting of established norms. That meant I was not a fan of Martin Luther King, Jr., who preached nonviolent resistance as the way to gains in racial equity. I was more aligned with Stokely Carmichael and Malcolm X and the Black Panthers, and I attended meetings of like-minded people. In fact, when Nate became President of the student body at Virginia State University and invited Martin Luther King, Jr., to speak, I opted not to attend the event because I feared I would be an embarrassment for my political views challenging the basic premise of nonviolent protest producing change. Nate is pictured with MLK from this visit in Fig. 4.1. Ultimately, I have come to understand that change is achieved more effectively from within the system than outside of it, although external pressure is often a necessary ingredient before internal activism can be appreciated. I guess that means that I now espouse a view midway between that of Martin Luther King, Jr., and Malcolm X.

In the latter part of my undergraduate degree program, I was encouraged to apply for internship opportunities. Recalling the problems I had getting accepted by study groups, I applied for every internship I could find. I was accepted to the summer Internship program at Bell Laboratories, the research arm of AT&T, and opted to accept this invitation. I was assigned to the Acoustics Department at the Murray Hill, New Jersey facility, even though I had no background in acoustics. I figured this would work out because it would be just like optics, which I had learned about, but with the signals in a very different frequency range. I was assigned to a group of scientists who were investigating human hearing. They were trying to determine the integration time of the ear. They would present a sharp click, a pause, and then another sharp click. The experiments involved varying the duration of the pause to determine the shortest pause at which people could hear two clicks instead of a single event.

Fig. 4.1 Nate with MLK at Virginia State University

The headphones the subjects wore were like condenser microphones run in reverse to produce sound rather than record it. In these sound drivers, a thin metal foil was suspended above a fixed metal plate and a DC voltage applied between the two. By adding an AC voltage of the click, pause, and second click, the foil would be expected to vibrate and generate a matching sound. Unfortunately, most subjects couldn't hear the click at all. I was asked to find an alternative sound driver.

I started the way all research begins—I went to the library to see if I could learn from work done by scientists preceding me. I found an article from a German scientist about a headphone built using a polymer film coated with metal instead of a foil comprised entirely of metal. I had the machine shop build something along the lines described in the paper. The driver needed to be hooked to a 500 V DC battery, so care needed to be taken not to shock the users, but the result worked beautifully. I was told that my accomplishment had saved the experiment, and I was invited to return the following summer.

This success as a Bell Labs intern was a turning point in my life. It was my first foray into research and the practical application of the science I'd been learning. It was exciting to solve a problem and to be a science contributor. Importantly, I had been judged on what I accomplished and not the color of my skin or the curliness of my hair. I had found my dream job. I returned to Temple University to complete my degree and hopefully, to find a job like those scientists at Bell Labs had.

In November, just a few months after completing my internship, I received a call from my Bell Labs bosses that the drivers had stopped working. I went back to the paper by the German scientist and, upon rereading it, found that they too had seen their headphones stop working, but had no idea why. I said I'd return at the semester end to see what I could figure out.

I spent my term break trying to determine what was happening with the drivers I had built. One of the first things I tried was to disconnect the batteries, expecting that the drivers would produce very little if any sound in this configuration. Imagine my surprise when the drivers instead sang beautifully. While this solved the immediate problem, I couldn't explain why this had happened. Even when I short-circuited the capacitor formed by the fixed and moving metal plates, the drivers worked! I had discovered electrets—materials, often polymers, that are permanently polarized. The batteries had polarized them.

My success in not only building new drivers, but then also fixing them, earned me an offer of employment in the Acoustics Research Department of Bell Labs at Murray Hill. I opted to leave school and accept this job rather than finish my degree. It was among the best decisions of my life.

5

Marriage and Bell Labs

Jessie Louisa Moss was born in 1937 to Mabel (Creft) Moss, a teacher, and Harold B. Moss, a carpenter, of Monroe, North Carolina. She was delivered by her grandfather, Hubert H. Creft, Sr., the well-respected local doctor and pharmacist. This baby, fully grown many years later, became my wife.

Dr. Hubert Henry Creft, Sr., the family patriarch, was born in Saint Georges, Grenada, British West Indies in 1885. He was orphaned when merely 12 years old and moved to the United States in June, 1905. Dr. Creft's wife was Lucretia (Christmas) Creft and their marriage produced three children: Mabel Creft Moss, born 1913, the mother of Jessie Louisa Moss, Dr. Hubert Henry Creft, Jr., born 1914, and Frank C. Creft, born 1915. Dr. Creft, Sr. trained first as a pharmacist and later as a doctor. Early on, he made house calls in a horse and buggy. It is thought that People's Drug Store of Monroe, North Carolina, created exclusively for people of color, was established by him and he occasionally served as their pharmacist.

As one of the only doctors for people of color in Monroe, Dr. Creft, Sr. was often called upon to certify the death of people who died in tragic accidents. On one such occasion, he was called in to the hospital from home to certify a death and found, when he removed the sheet, his younger son, Frank, lying on the stretcher. I know that this event scarred him and affected the family greatly.

Dr. Creft, Sr's wife was very light skinned and is thought to have had a white ancestor in the past. Dr. Creft was also light skinned. The couple faced blatant colorism. At one point they were mugged, an action attributed to her light skin tones and their relative affluence while living in a Black community.

J. E. West, I. J. Busch-Vishniac, *Mic Drop*, https://doi.org/10.1007/978-3-032-20922-1_5

In 1960, Dr. Creft, Sr. was named Doctor of the Year by the Old North State Medical Society for "meritorious service to the profession and to the professional organization of which he is a member" in an event held on the campus of North Carolina A&T. Dr. Creft, Sr. worked as long as he was able. When his hearing made it difficult to listen through a stethoscope, he had me rig something that would help him. It was a clunky, heavy stethoscope with a microphone and earphone, but it did the trick. This was the first of my three times working with stethoscopes. Dr. Creft, Sr. passed on in 1965 and in 1978, Creft Park in Monroe was dedicated in his memory.

The Moss family and the family of Hubert Creft, Jr. lived in two houses behind the large home of Hubert Creft, Sr., creating a family compound where children interacted regularly with their cousins, aunts and uncles, parents, and grandparents. It was in this family compound, similar to that of my youth, that Louisa, as she came to be known, grew up. The family is pictured in Fig. 5.1. Unlike my family, which was comfortable but lacked funds for luxuries, Louisa's was well off financially.

Another one of the differences between my family and that of Louisa, was in how they dealt with the indignities of racism. My family chose to tolerate it and avoid actions that would irritate racist white people. Louisa's family

Fig. 5.1 The Moss family. On the sofa, from left to right: Harold Moss, Mabel Moss, Dr. Hubert Creft. Behind the sofa, from left to right: Harold Moss, Jr., Louisa Moss, Henry Moss, and Annabelle Creft (wife Dr. Hubert Creft)

generally chose a more confrontational stance. Two stories illustrate this difference. Dr. Hubert Creft, Jr., the son who followed in his father's footsteps and became a doctor with a practice in High Point, North Carolina, was one of three Black doctors who, in 1954, forced the City of High Point to reconsider its long-standing policy of racial segregation of recreational facilities by insisting on playing golf at the segregated Blair Park municipal course. This action served as a catalyst for others to challenge racist policies in High Point and in 2022, the Museum and the City of High Point Parks and Recreation Department along with the North Carolina African American Heritage Commission held an unveiling ceremony at Blair Park for the historical marker honoring the 1954 golf course protest.

Very much later, when Louisa and I went to visit her family, Dr. Creft, Jr's younger kids and ours went shopping for Christmas gifts. We found ourselves in a store when Hubert's young daughter, Alice Gwendolyn, announced she needed a bathroom. We flagged down a salesperson and asked for directions. We were told there were no bathrooms for "niggers" available. Hubert looked at his daughter and said he wouldn't have her suffer because of the ignorance of a salesperson and pulled down her pants and directed her to pee right there, leaving the salesperson fuming.

Mabel Moss worked as a schoolteacher in Monroe. Harold Moss was said to be a carpenter, but my recollection is that he spent a lot of time just hanging out with his friends. His mother, Mama Moss, was an amazing cook and made wine which I enjoyed. Louisa didn't speak much of her father, but she was very close to her mother and to her brother, Hubert.

Louisa was an incredibly good student. There was truly nothing she couldn't do if she put her mind to it. She read voraciously as a child and an adult. Louisa also was a classically trained and very accomplished pianist. She graduated from high school at 15 and opted to attend Fisk University in Nashville, Tennessee for a bachelor's degree in psychology. Fisk University was founded just after the Civil War with an aim of being open to all and providing a high-quality education. Its first students varied greatly in age but shared a history of being enslaved, of enduring poverty, and of having a huge thirst for education. By 1871, just 4 years after Fisk University was incorporated, funding of the school had become a major obstacle. The Fisk Jubilee Singers were formed and traveled the world to attract donations from white donors to keep the school going. Today, the Fisk Jubilee Singers are a Grammy recipient group, and the tradition of electrifying and inspiring singing continues. While at Fisk, Louisa joined the Fisk Jubilee Singers. She graduated with her psychology degree at the age of 19.

Louisa entered the University of Pennsylvania to pursue a PhD in psychology and it was while she was there that I met her. There weren't a large number of Black people in the medical arena in Philadelphia at that time, and I think we met at one of the social events of this group. I knew immediately that here was a woman much smarter than me and we had wonderful conversations. Louisa was bubbly and fun to be with. She saw the beauty in everything and helped me to see it. With each of us being consumed at that time by our studies, we saw each other infrequently for a year before things became serious between us.

Louisa quit her graduate program about 6 months before we married. She had come to see many of the research experiments in her field as unethical for how the subjects were treated. She decided instead to move to social work, and to focus on actions to help people in need.

At the point that Louisa and I married, in December 1957, I was an intern at Bell Labs and on my way to a permanent position there. We were happy and Louisa's family and mine made us feel very comfortable. We enjoyed visiting Louisa's family as often as possible. We moved into a house in Plainfield, NJ. It wasn't long before we started having children. Our first, Melanie, was born in October 1958. She was delivered by her great grandfather, Dr. Creft, Sr. Enid Johns, Louisa's friend for life since their meeting at Fisk, named the baby Melanie and served as her godmother. Louisa named Enid's first child Beth and served as her godmother in symmetry with Enid's role for Melanie. Enid was also from Farmville and had been part of the team of people that spoke at the protest of the quality of the schools along with Edwilda.

Two and a half years after Melanie was born Louisa and I had our second daughter, Laurie, in 1961. Like Melanie, Laurie was delivered by her great grandfather. With two youngsters at home, Louisa took time off from her social work job to be at home caring for the girls. Six years after the birth of Laurie, Jay was born. Enid referred to him as my little distraction and this was not a misstatement.

The three kids are similar but different from one another. I think of Melanie as my feisty child. She has always loved a good argument, and she challenges me regularly. By contrast, Laurie is calm and introspective. Even today, Laurie's visits are the most calming part of my life. Jay was a typical, rambunctious boy who ended up in the emergency room with fair regularity.

My kids describe their early years as "magic." Louisa and I were in agreement that the kids should be exposed to a wide range of experiences and be allowed to chart their own course forward. Our job was to make sure they were safe as they tried new things and that they knew they had our support. I taught them how to swim and I also took them ice skating. Louisa regularly threw parties for the kids and showed them the beauty surrounding them. She made all holidays spectacular and ensured they were exposed to culturally enriching experiences. To the extent that there was a fun parent, it was Louisa. I was the one who reminded the kids to do chores. I also took them food shopping with me, even though shopping with kids can require a lot of repetitions of the phrase, "No, we don't need that".

Where we lived in Plainfield was an ethnically mixed community. Plainfield had been founded by Quakers but had subsequently become a bedroom community for people working in New York City. By the time we moved there, Plainfield had developed some of its own industrial base and there was an influx of workers to support it. We lived in a working-class neighborhood on the east side with all sorts of Black people and immigrants as neighbors. There was an Italian bakery across the street from us, and we woke up each morning to the aroma of fresh baked bread. The bakery made wonderful pizza which became our regular Friday dinner. Our kids were often at that shop purchasing cookies. There was also a playground nearby and the kids spent much of their time outdoors.

Our house was a Cape Cod style. The children's bedrooms were on the second floor and the master bedroom on the first floor. A family friend built shelves and closets into the house for us and helped me maintain it.

Louisa and I were living the good life in Plainfield. We always had enough money to do whatever we wanted to do, and the kids lacked for nothing important. I've never been great at keeping track of my money, but I know it was sufficient in those days to let us be very comfortable and to splurge as we wanted. That's always been my measure of success financially. Do I have enough to do what I want? In Plainfield, the answer was pretty much yes, although we didn't take pains to save money for the future.

Plainfield also was convenient for me in terms of getting to the lab. I would regularly bike to work up Diamond Hill Road. It was a brutal climb, but it was my way of getting exercise without wasting time. If I stayed late and it got dark, Louisa would pick me up in the car.

Life at work was going well. My internship at Bell Labs had opened my eyes to the opportunities in corporate research. I had considered a few companies for permanent employment but ultimately accepted the offer from Bell Labs because I saw some people already there who looked like me. I was hired by department head Ed David, who warned me that everyone at Bell Labs got picked on for one reason or another, and that I should let it roll off me. Ed David was an electrical engineer who worked at Bell Labs until 1970, when he became the Director of the Office of Science and Technology under

President Richard Nixon. His warning was useful, as it let me prepare for being part of the Bell Labs crew.

While there were many people at Bell Labs who wanted to see me succeed, it was W. Lincoln (Link) Hawkins I turned to for mentorship. Link was the grandson of a slave and was a stellar chemist. He was a graduate of Rensselaer Polytechnic Institute with a BS in chemical engineering, where he was one of only two Black students at the school at the time. He earned a master's degree in chemistry from Howard University and a PhD in chemistry from McGill University. In 1942, Link became the first Black person on the technical staff of Bell Labs. His most well-known accomplishment at Bell was the development of new insulating materials for telephone cables. Prior to his work, cables had been first wrapped in fiber and then in very expensive and heavy lead. As plastics were developed, the aim was to use them as a replacement cable insulator. The problem was that many of the plastics were reactive to elements in the environment. They would become brittle in the sun and break. Eventually Link and his collaborator, Vincent Lanza, invented a coating that could withstand a wide temperature range and last up to 70 years without deterioration. It was this work that was largely responsible for the expansion of the telephone network into rural areas, as the lowered cost made the expansion a reasonable expense.

Link agreed to be my mentor. He said, "I'll teach you what I know and advise you the best I can based on my experience." I listened carefully to his advice and Link said he was glad for this because his son, about the same age as me, didn't heed his advice.

My immediate supervisor at Bell Labs was David Berkley. I am forever indebted to him because he worked hard to give me a warning before I got into too much trouble. David didn't always manage to keep me safe, but he tried. My department head became Manfred Schroeder, an architectural acoustics scholar, after Ed David moved up in the Bell Labs management heirarchy. Later in my career, Jim Flanagan, a speech recognition expert, became my department head and Max Matthews became the center director, Jim Flanagan's boss. Max was considered one of the founders of electronic musical instruments. He was a violin player who usually played second violin. Max would walk the halls whistling the music from some piece he had on his mind. This was great because we all knew when he was around and likely to interrupt us to see what we were doing.

I recall being worried when I learned that Jim Flanagan, a speech generation expert, would take over as department head. I responded to the news of this change by visiting Bill Schlichter, a mentor and friend. I told him I thought my career was in danger because I couldn't expect a southern gentleman from Mississippi to see beyond the color of my skin to my accomplishments. Bill told me to give Jim a chance, but I think he must have spoken to Jim, because Jim's behavior was diametrically opposed to what I expected. He was very friendly and supportive. He regularly went to bat for me. He was one of the few colleagues who invited me to his home for dinner and Louisa and I enjoyed the company of Jim and his wife, Mildred.

Having started my time at Bell Labs working with electret materials, my first research project was to delve deeper into this phenomenon of materials that charge up (polarize) and retain that charge. I wanted to get a fundamental physical understanding of the process the materials were undergoing. Fortunately for me, I was able to convince Gerhard Sessler, who arrived at Bell Labs about the same time I did, that this was an interesting question, and we began what would be a decades-long collaboration. It helped that Gerhard and I had offices and labs next to one another. Gerhard Sessler was born in Germany just 5 days after my own birth. He studied physics there, graduating with a PhD from the University of Gottingen in 1959. Upon graduation, he joined Bell Labs. Gerhard was married to Renate Sessler and our families would occasionally socialize.

Both Gerhard and I saw that electrets might be very useful in acoustic sensors, and we started running tests to determine what materials behaved as electrets, how one might initially charge the materials, and how long the polarization would remain. We looked at various means of depolarizing the material and studied material structures using an electron microscope. In short order we became experts on electrets.

Gerhard and I were not the first to study electrets. They had been known for years but not thought to be terribly interesting or useful. Famously, Michael Faraday, a well-known physicist for whom the unit of capacitance is named, had said their only use would be to teach electrostatics to students. Well, Gerhard and I proved that statement wrong. Beginning in 1962 we published a series of papers which described a new microphone using electrets as the sensing surface and were granted a patent for this work. In its simplest form, an electret material such as fluoroethylene propylene (Teflon) is coated on one side with a very thin metal layer. The electret is permanently charged and then suspended over a rigid backplate so that the plate and the metal coating form the two plates of a capacitor, with air and the electret material between the plates. Sound striking the electret foil causes it to move which changes the distance between the two plates, which, when a static voltage is applied between the plates, produces a dynamic voltage that mirrors the foil motion. It is the charge on the electret that provides the static, sometimes called bias, voltage. The dynamic voltage is the signal that mirrors the sound pressure.

The microphone in use in telephones worldwide at that time was the carbon button microphone. Although typically credited to Alexander Graham Bell, it was invented by Granville T. Woods and purchased by Bell. Woods was a Black, prolific inventor. Among his inventions were a railroad telegraph that transmitted messages between moving trains, the third rail to power trains using electricity, and an overhead electrical conducting system for trolleys. The carbon button microphone operated by placing carbon granules between two plates. As sound moved the thinner, outward-facing plate, the carbon granules would be compressed and the electrical resistance between the plates would change in a pattern matching the sound.

Because I worked at Bell Labs, an AT&T subsidiary, I learned a story about the carbon button microphones that is not widely known. AT&T made these microphones in house. Part of the process involved roasting the carbon granules in ovens. At some point, Ma Bell (as AT&T was colloquially called) decided to replace their ovens because they were old and inefficient, but when they did so, the carbon button microphones produced didn't work. After searching for an explanation, it was quickly concluded that the oils that had built up in the old ovens were apparently a necessary part of the activation process. The new ovens were removed and the old ovens moved back into the manufacturing plant.

A more advanced microphone, invented by Edward Wente of Bell Labs, was available at this time but not in use in telephones. The Wente microphone is the condenser microphone. It uses a plate that moves in response to sound suspended over a fixed plate. The changing gap between the two changes the capacitance. In essence it is the same process as the electret microphone except that it requires the application of a bias with a battery. In the electret, the bias is provided by the polarized polymer.

The electret microphone Gerhard and I invented had many advantages over the microphones that existed at the time. They could operate with very little added power—just enough to amplify the signal. They were inherently very linear, so doubling the intensity of the sound would double the signal strength. They were very low in distortion, so the signals produced were very clean. Finally, they were inexpensive compared to the existing microphones. Our department head, Jim Flanagan, was very impressed. It was clear that we had the key to replacing the round, expensive, and very nonlinear carbon button microphone in the telephone handset. Our solution would cost less, weigh less, and perform better.

The Bell Labs policy was to patent all inventions, so Gerhard and I filed a patent application in 1962. It was granted in 1964. It was the first of the more than 250 US and foreign patents I have been issued in my career—none of which have affected my finances significantly and only a few of which have been commercialized. Although the courts had ruled that patents for work done as a part of your employment at a company were owned by the companies, Bell Labs always paid a very modest bonus for a patent to make it clear that they owned the invention. I believe that Gerhard and I shared $1000 for this first patent. It was also Bell Labs policy that a paper could not be sent in for publication without first clearing the patent office to ensure that all potentially commercial work was protected. This accounts for some of the patents I have in my name. I couldn't send in the paper without the patent application going in first.

Jim Flanagan took our invention to the product development people at AT&T and was told that they weren't interested. It was a frustrating time for all concerned. Jim Flanagan was thrilled at the success we had had, but sad that we would now leave, of course, to commercialize the product ourselves since Bell had waived their rights to it.

Gerhard and I didn't take long discussing the option of starting a company. It wasn't for us. We were both researchers and happy in our circumstances at Bell Labs. Leaving to start a company felt like learning an entirely new discipline in a foreign country with a language we didn't speak. We passed up the opportunity, deciding instead to keep working to improve the electret device and to develop a greater understanding of the fundamental physics behind its operation. Viewed with 20:20 hindsight, I wonder how things might have panned out if we had taken up the challenge of commercializing the electret microphone. Ultimately, many companies arose to manufacture and sell inexpensive electret microphones. Among these Primo and ECMIC were the largest producers. Starting about 1966 and going until 2014 or so, the electret microphone was the most produced microphone in the world, peaking at more than two billion units sold every year.

In these early years at Bell Labs, I learned that the people in research would pretty much always be thinking about their work. I would eat at the Bell Labs cafeteria with the same group of people most days. The cafeteria had paper placemats at every seat. I learned that the first thing you did upon sitting down was to remove that placemat and turn it over so that the blank side was facing up and you could write on it. On one occasion the lunch ended with my placemat having about ten equations and a few words on it here and there. I took the placemat to my department head and explained I needed $5000 to test the theory. Based on that placemat, I got the money I needed for the lab tests.

One of the nice things about Bell Labs was that it was normal to work on multiple projects simultaneously. I found that useful because it gave me a way to work in the lab on one project while thinking about a challenge presented by a different project. I've always worked best with busy hands and this multi-project approach was perfect for me. Having busy hands seems to free my brain to think about technical issues unencumbered. Dreaming does the same to my brain. To this day, I keep a notebook by my bedside so that I can jot down an idea that might come to me in a dream as I sleep.

In these early days at Bell Labs, I developed a surprising friendship with Ben Logan. We came from totally different cultures, but Ben played bluegrass music on his fiddle every day at noon, and this was the music I had grown up with. It allowed us to form a bond of mutual appreciation. While I enjoyed Ben's playing, I didn't realize at first just how accomplished a musician he was. At one point Ben disappeared for about a month. This was not unusual at Bell Labs, but a staff member asked me if I'd seen Ben since she had two of his paychecks. I said I hadn't seen Ben, but that night when I turned on the TV to watch the Grand Ole Opry, who should I see but Ben Logan performing on stage on his fiddle. That is how I came to know that Ben, known to the music world as Tex Logan, was perhaps more famous as a musician than as a scientific researcher.

Early on in my time at Bell Labs, I was asked to work with Manfred Schroeder, Bishnu Atal, and Gerhard to renovate the acoustics at Philharmonic Hall in New York City. At the time Manfred was a department head, and he led the project. Bishnu and I had shared an office when he first arrived at Bell Labs. He talked about what it was like living in India under the rule of Britain, with rules imposed on them from outside their culture. That perspective allowed Bishnu and me to share common ground.

Philharmonic Hall had originally been designed by Leo Beranek, one of the founders of the acoustical consulting firm, Bolt, Beranek, and Newman. Leo had installed tiltable reflectors he called clouds high in the hall to reflect sound back to the audience. The problem that had been discovered is that these clouds tended to trap the low frequency sound high up in the hall. Critics had panned the sound as lacking in the low frequencies, except for a reviewer from the *New York Times* who had judged the hall after being seated in the balcony for the performance.

A decision had been made to renovate the hall and bids were accepted. Bell Labs could not bid on the job given its corporate status, and the job went to Vern Knudsen. Vern was not local, so he contracted with Bell Labs to consult on the project and make measurements. Only Manfred knew anything about architectural acoustics at the start of the project, and Bishnu, Gerhard, and I saw this as a chance to learn.

We began the project by traveling to a few symphonic halls and listening to them. We found that each had a unique sound—some good, some great, some not so good. The team developed a computer model of room acoustics that permitted us to listen to what we believed the hall would sound like given its architecture and the materials used. The program was so large it had to run on an IBM 360, the largest computing power available at the time, and we would run it on weekends so as not to prevent others from using the computer during working hours. Now, you'd probably be able to run that same code on a mobile phone. Using this computer model, we produced renovation designs for Knudsen's company. Once they were satisfied with the designs, Knudsen made the arrangements for the renovation to incorporate the design. We would be needed again to run some measurements at various stages of the reconstruction and completion.

Manfred Schroeder also got involved with Gerhard and me on our work on directional microphones, i.e. microphones that responded preferably to sound coming from specific directions. Manfred was a very interesting physicist originally from Germany. It was impossible to go to a brainstorming session with him and not come away with new ideas. Manfred was already well known for his work on the voice-excited vocoder, a speech analyzer. He went on to become very well known for his work on linear predictive coding, a way of representing the frequency content of a sound signal in compressed form, and for many insights regarding architectural acoustics.

Interestingly, the brainstorming sessions with Manfred, Gerhard, Bishnu, and me very often produced three distinct views. Manfred and Gerhard tended to agree on ideas. Bishnu and I would usually have a different perspective on the topic and wouldn't even agree with one another. So, four people and three different perspectives. I have always believed this was due to our distinct cultural upbringings. I am certain that how you are raised impacts how you ask questions and think about the world. This partnership seemed to demonstrate that. It also is a prime example of how important it can be to have a diverse research group to ensure that the issues are viewed from many perspectives. Because I have tended to be the only Black person in a technical brainstorming session, I have more often than not brought in a different perspective, and generally that perspective has affected research directions as long as the team was willing to let me be heard.

In these early days at Bell Labs, it was good that I remembered what Ed David told me when I was hired. The technical staff in my department tended to gather every day at about 10 am for coffee and discussion. There was talk going on that I just didn't understand. I remember they kept referring to somebody as a DJB and I had no clue what that meant, although it clearly seemed pejorative from the tone of the comment. Never being shy about learning, I asked people "What is a DJB?" but nobody would answer until finally Ben Logan told me it was a "Dirty Jewish Bastard." My response was, "Well, I know what you are calling me." I was, after all, the first and only Black person in the division.

Interestingly, there was a strict divide between scientific and nonscientific issues in personal interactions. I might be called a nasty name, and my political stances might be questioned, but my scientific work was valued and respected. I chalk that up to the fact that scientific questions have a single correct answer and many, many wrong answers. There is nothing in between. If your theory stands up to testing in the lab, then you have the right answer regardless of your gender or skin color. It was an uncomfortable situation to learn that I was called a nasty name but oddly, I was unfazed by this as there seemed to be a clear separation between social relations and work relationships. My work was going well, and I was valued by these same folks who were calling others, and me, nasty names. I let that be enough and did as Ed David suggested—I let it just roll off me.

Another early incident related to door decorating. At Bell Labs, office doors were traditionally kept unlocked and often open. People were encouraged to just walk in and talk about work-related topics. Many of the staff decorated their doors with personal or work-related images. I chose to put a Black liberation flag on my door, but it got vandalized and no other decorations on doors were touched. I asked around and got told that Mark Gardner had scraped the flag off my door. He had a large US flag on his door, so one evening, I took his flag off and epoxied it hard to his door upside down. I remember that HR contacted me after this and asked if I really wanted to get involved in a fight. I didn't tell them directly how I felt, but I felt vindicated at the time. Mark messed with me, and I got back at him. He destroyed his US flag getting it off his door and that felt like it made us even. I told HR I would leave things as they were and not pursue it further if he didn't. I put another Black liberation flag on my door. This time it remained untouched.

I also found a group of guys at Bell Labs interested in racing cars. This had become a passion of mine and I was determined to succeed as a race car driver. Newly flush financially, I purchased an old Porsche 365 and rebuilt it. I loved the power of that car and regularly tested it on the roads. This went well until I had a stupid accident that totaled the car. I was on an entrance ramp to a highway near the Delaware Water Gap. There was a slow-moving truck in front of me with no one coming up behind me, so I figured I could pull around the truck and zip onto the highway. The truck, for reasons I still don't understand, decided to hit the brakes while I was just starting to pass him, and I ended up slamming into his rear. Luckily, although the car was a mess, I got away with only minor bruises. It did, however, put an end to my race car driving for a long while. Fortunately, at least in my mind, I was asked much later in my career to help with a communication problem between the pit crew and a race car Formula 1 driver owing to the noise in the car. I made them a new microphone, one that was highly selective for sound coming from the speaker's mouth, and that helped. In gratitude, the team let me come to Watkins Glen, New York and race the BMW on the track. I was so jazzed about this that I remember buying race car driving shoes, with pointed toes that make it easier to shift quickly from accelerator to brake with your right foot. However, Watkins Glen was the end of my racing career. My family wasn't comfortable with me pursuing such a dangerous sport.

At work, it did not go unnoticed that I was collaborating in these early years with scientists born outside of the US. In particular, there were comments about my willingness to collaborate with the Nazis, referring to Gerhard and Manfred, my German-born colleagues. I went to see my mentor, Link Hawkins, and asked for advice. He told me that in order to collaborate with white, US-born scientists, I would likely have to give them more credit on a project than they deserved. I kept that thought in mind as I sought additional collaborations and projects. It took some time, but I did manage to accumulate a host of US-born collaborators over the years.

Overall, I was doing well and happy by the early 1960s. I had a loving wife and children, lived in a comfortable community, and thoroughly enjoyed my research work at Bell Labs.

6

My Plainfield Community

The year 1967 was tumultuous across the US, as race riots broke out and cities burned. Despite the Civil Rights Act of 1964, segregation in housing and education was still the norm, leading to frustration at the slow pace of change. In Plainfield, where my family was living, projects to create housing made segregation worse, banks continued to have discriminatory lending practices, and the high school had two education tracks for students: one for (white) college-bound students and one for (Black) students headed straight to the workforce. Newark, New Jersey erupted in mid-July and Plainfield, merely 18 miles away, followed suit a day later. In just a few days, the riot was over, but a Plainfield police officer had been killed, 23 protesters had been shot, and 167 people had been arrested. Damage to the city was estimated to be more than $700,000, an amount that would exceed $5 million in 2025 dollars. This was the environment my family faced in the late 1960s.

Coming out of the racial unrest of 1967, a group of Black scientists who lived in Plainfield put their minds to what they could do to help restore order and improve the lives of people in their community. We came up with the idea to start a science center focused on giving children a way to spend their time learning rather than causing or being consumed by trouble. Besides me, the group included Jim Mitchell, Lloyd Sheppard, and Doug Osheroff. We pooled our resources and rented a storefront in downtown Plainfield on West Third Street, in a former appliance shop that suffered damage in the riot. This location put us directly across the street from the community swimming pool. In short order, the Plainfield Science Center became so popular that we had kids on the sidewalk waiting to come in and interact with our displays. Doug

J. E. West, I. J. Busch-Vishniac, *Mic Drop*, https://doi.org/10.1007/978-3-032-20922-1_6

later received a Nobel Prize in Physics and contributed some of his prize money to keeping the center open.

Over the next years we built up activities and programs in the Plainfield Science Center and started applying for grants to support it. We found a director and brought in interns to help run programs. We created a parent organization called the Plainfield Science Education Coalition to engage parents in our activities and engender their support. By 1971 we had a National Science Foundation (NSF) grant to help support the center, we were awarded some state funds, and we hired Al Waller as the director and only full-time employee. The center by then was providing lab experiments, research projects, group and personal tutoring, and trips to various sites including museums, zoos, nature preserves, weather stations, and agricultural centers. It also ran a summer camp. We bought a house on Seventh Street and moved the operation. As reported in a *New York Times* article about the center, we had more than 500 children, aged 5–17, participating in the center's programs. My children were among them. In fact, both Jay and Melanie worked at the center at some point—Melanie as a math instructor and Jay as a general helper.

Largely because of the race riots that roiled the US in 1967 and 1968, there was a political backlash against efforts to focus on improving the lives of people in poor communities, particularly Black-dominated communities. To me, this has always seemed like the opposite of the appropriate reaction. The riots were fueled by frustration that people were locked out of the same opportunities provided to their white counterparts. A reasonable response should have been to unlock those doors, but instead, a large part of the country decided that anyone who participated in riots must be an animal and should be treated like one. The Science Center lost some of its state funding for a decade or so, ultimately winning it back, but we didn't let the lack of funding curtail our activities. The Plainfield Science Center lasted for over 20 years.

There were many stories from the Plainfield Science Center—people it saved, people who were beyond saving as the streets consumed them. One story was of a youngster I taught to play chess. He came to the center one day and asked if he could have a chess board. I gave him a board to take home and the next week he wanted another, so thinking maybe he lost or ruined the first, I got him another. But then he returned and asked for yet another board. I asked what he was doing with all these chess boards, and he told me he was lining them up and teaching other kids to play chess, so he'd have someone to play with. Unfortunately, this story doesn't have a happy ending. Years later this same young man died of a heroin overdose.

Looking back on this time period, I would say the 1960s and 1970s were the hard part of being Black. I was convinced that anybody white was against me, especially if they were from the South.

During these years, life at home was still good. In addition to the kids being involved in the Plainfield Science Center, I would take them with me on occasion to various places I needed to visit. I remember, for instance, that the kids were with me when I returned to Philharmonic Hall to make measurements of the acoustics after the renovation. I asked the security people to keep an eye on my children but told the kids they were on their own for a bit as I worked. This was part of my constant attempt to let them be independent, confident, and inquisitive in the world. It also let my children spend time with me and see what sort of activities I engaged in at work. On one outing, Jay ended up on his own at the Guggenheim Museum. The circular staircase caught his eye, and he started dropping pennies to see how long it took them to hit the bottom. I was summoned and we put a stop to this as it was potentially dangerous.

I also made sure that the kids saw my friends and colleagues so that they saw successful Black men and women and not just the folks hanging out at street corners. I was intent on destroying the myth that most Black men were in jail for illegal activity and those who weren't in jail weren't worth the salt in their bodies. There were constantly doctors, lawyers, judges, and all sorts of professionals coming to the house. Coupled with Nate's faculty position at Howard University, the children certainly had appropriate exposure to role models. Many of the visitors were successful musicians. Among them were Jeffrey Bowen, a recording artist and record producer known for his work on Motown records, and his wife Bonnie Pointer, one of the Pointer Sisters. Bernie Worrell, keyboardist for the Funkadelics and talented composer also made frequent appearances at our home. The only downside to this constant presence of successful and impactful people, was that at least one of the kids felt it set a bar too high to live up to. I've only come to understand this recently, so I couldn't correct that misapprehension when it happened.

The house we lived in had originally had an above ground pool. I removed it and converted the area to a garden. From my early days, when almost all the food we ate was grown on the farm, I knew quite a bit about gardening, and I taught the kids. Laurie especially absorbed those lessons and has ever since sought to live where she can have a large garden.

As is normal, the kids would sometimes be best friends and sometimes be enemies. Jay, as the youngest, was occasionally picked on by his older sisters—by Laurie in particular. Melanie would sometimes rush to his defense. Fortunately, they have overcome this normal sibling rivalry and fighting and are now all good friends.

Because I was working long and late hours, I didn't see the kids as often as I would like. They were normally already asleep by the time I got home and frequently left for school before I woke up. Weekends was for us to bond, though. We'd start with chores and cleaning up on Saturday morning, and then we'd all go grocery shopping. I'd try to add an outing such as a trip to a flea market or a small town in the area to the agenda.

At this point Louisa was working as a student counselor at Rutgers University. The students loved her, and she thrived in her job. There were often Rutgers students around the house mingling with our family. But a troubling pattern was emerging. While Louisa could be bubbly, positive, and focused, she was also prone to depression. At times she would just go into the bedroom and lock the door so she couldn't be bothered, sometimes for quite a bit of time. On one occasion, we had to call someone to remove the door from its hinges so we could get to her and get her help. Louisa's depression had a huge impact on the entire family. It put a burden on me that I was reluctant to carry, and it soured our relationship.

Louisa also participated actively in the local Episcopal Church, where she served as music director. She regularly took the kids with her to Sunday services, but I was unwilling to go, having given up on religion much earlier. The church music became part of the repertoire of music constantly around in our home. The children took music lessons and even listened (reluctantly) to opera, as Louisa sought to broaden their appreciation of the arts.

While life at home was challenging at times, professional life at Bell Labs continued to be rewarding. Gerhard and I were making good progress on understanding the fundamental physics behind electrets.

7

Understanding Electrets

Life at work was progressing nicely. An important aspect of being a scientist is presenting your results to others in the field, hopefully leading to fruitful discussions. Professional societies play a key role in enabling discussion by hosting conferences, so it was clear I needed to join a society or two. There were two professional societies it made sense for me to join—the Acoustical Society of America (ASA) and the Institute of Electrical and Electronics Engineers (IEEE). It was Manfred who introduced me to the ASA and helped me with my first talk there. It went well, but I was asked some funky questions after my talk by acousticians in attendance. Notably, I was asked what specifically my contribution had been. This sounded an awful lot like I was being told that surely a young Black man couldn't have done this work. The leadership of the ASA, by contrast, were effusive in welcoming me to the society. Cyril Harris was society president at the time, and he made a point of finding me and telling me how delighted they were to have me there. I'm guessing I was easy to find since the society then was almost entirely white, and unfortunately still is.

This issue of assuming that my work must have been either the work of others or coming from the ideas of others was a recurring theme at work. I recall that my research was often challenged at Bell Labs. No doubt about it. There were a few people who said "Well, you got that idea from me." But I had a counter argument that was absolute, which is one of the reasons that I like science, more than other disciplines, because it relies on documentation. I would ask them for the report or paper or notes where they outlined what I had done, and they couldn't produce anything. We were all required to keep

J. E. West, I. J. Busch-Vishniac, *Mic Drop*, https://doi.org/10.1007/978-3-032-20922-1_7

lab notebooks and when theirs showed no signs of thinking about an idea I had presented, it was clear they were simply trying to undermine me.

At Bell Labs, most people had seemingly gotten more comfortable around me, and we would occasionally talk about racism and similar issues. One such conversation was with Henry Pollack, a well-known mathematician. He had gone on a study tour of Africa and found the students there to be very accomplished in mathematics for their grade level. The problem was that there was nothing to take them further once they finished a certain level of school. This disturbed Henry greatly and he talked with me about what these students would likely be capable of doing if only the resources were available to them. It is this issue of availability of resources that occupies my thoughts often. I often think about what life would have been had I had the opportunities that I deserved at the time I merited them. I have no way of predicting what that future might have held but considering what I've been able to accomplish despite not being given a fair opportunity, I may have been able to make extraordinary contributions beyond what I've accomplished.

As time progressed, I developed a civil and light-hearted means of dealing with low-level racist events. I had learned that humor often helped to make a point firmly but without aggression. As an example, I once had a visit from a mathematician from Kentucky. He had read about my work on the electret microphone and wanted to discuss it with me. He traveled to Murray Hill and when I picked him up at reception, to come to my office, he asked me, "Where's Dr. West?" I said I'd see if I could find him. "Oh, yeah, here I am." It hadn't occurred to my visitor that I might be Black. Nor had it occurred to him that the title "Dr." might be inappropriate, as Bell Labs research scientists were assumed to have PhDs. My visitor was embarrassed, and I dare say was unlikely to repeat such an event again, but he didn't feel attacked. We were able to continue with our meeting and to focus on technical matters.

Related to issues of diversity and equity, Bell Labs had a program we nicknamed "charm school." In this program, a small group of employees were taken to a remote location and given an opportunity to talk about what made them and their scientific work the same as others and different from others. It was an attempt to focus on diversity as a good characteristic of Bell Labs and to reinforce diversity through this forced socialization. I was invited to one of the charm school meetings to be held in Baltimore. I arrived at the hotel the same time the leader of the group was checking in, and I was told they had no room for me. They insisted there was no reservation. The group leader intervened and said, "He's part of our group. Give him a room." The hotel management still refused. They said they didn't have any rooms for Black people, although they used a different word to describe me. The group leader responded by telling everyone to go home. The meeting was cancelled. It reinforced my comfort at Bell Labs that the group leader acted as he did. He was clearly more shocked by this event than I. It was not my first time to be confronted with racial discrimination and it would not be my last. I guess charm school should have avoided Charm City as a venue.

Gerhard and I continued our work on electret sensors and published a long list of papers documenting what we found. We are shown together in a photo (Fig. 7.1) from that time. Among the items we invented was a touchpad for the telephone with a patent awarded in 1970. Work in this area had been suggested by Bob Wallace, a senior circuit engineer with Bell Labs, and well-known racist. This invention was picked up by the development folks at Bell Labs and eventually the rotary dial was replaced with the touchpad, but not the design we came up with. The touchpad is still used in phones, although the design has changed considerably as technology has advanced. It is faster than a rotary dial and less prone to breaking. It operates with each button sending a signal with a unique frequency that the system converts to a phone number for routing purposes.

In the early 1970s, Gerhard and I realized that we could use some additional thoughts on the behavior of electret materials. We were particularly interested in how to polarize the materials we were using. Corona charging, in which a material is exposed to a conductor at high voltage, seemed like the right approach but it was hard to control. We tried playing with a Vandergraff generator (the machine that museums have that will make your hair stand out as you touch it) as an alternative, and this led us to think about radiation and polymers. With David Berkley, my immediate supervisor, we also worked on using gamma radiation to polarize Teflon, but the results were uninspiring. That led us to Bernard Gross, an expert on electronic materials.

Bernard Gross was born in Stuttgart, Germany in 1904 but left for Brazil in 1933 as he judged his chances to contribute significantly to physics would be greater there. His original work was on cosmic rays, but in his long career he published papers on gamma rays as well as on dielectric materials. A dielectric is an electrical insulator that can be polarized by an applied electric field, so electrets are dielectric materials. At the time we contacted him, Bernard was Brazil's representative on the International Atomic Energy Agency but living in Vienna. Gerhard and I went to visit him to discuss what we were seeing with electret materials and what we were trying to accomplish. We spent our time with him forging what would be a lasting collaboration and designing experiments to test various theories. The visit was professionally wonderful, but we saw none of Vienna. We took all our meals at the little café near Bernard's home and spent all our time working.

Fig. 7.1 Gerhard Sessler (seated) and me in the lab, holding electret film. An unnamed person is holding a miniature electret microphone

Bernard was a theorist, and I am an experimentalist. Gerhard falls between the two of us, with comfort in both theory and experiments. As our collaboration aged, Bernard was able to produce predictions from his models that matched experiments well, with one exception. Bernard presented me, at one point in our collaboration, with results I should expect to see from an experiment we had agreed to do. Try as I might, I couldn't get the results in the lab to agree with his model. Forlornly, I told Bernard that the experimental tests, which produced the same results over and over again, did not agree with his model predictions. It was then that Bernard surprised me. He told me that he had intentionally given me erroneous predictions to make sure that I was accurately reporting what I measured. I was angry at first, but Bernard explained that our agreement between his models and the lab results was so good that he had worried that perhaps the lab results had been made up. He was making sure that this was not the case.

In 1973, I published my first paper with an American-born coauthor. More than a decade had passed since I had joined Bell Labs, but this was the first time I could engage American colleagues in the problems I was working on. The issue at hand was very interesting. In 1968, Sony produced the first commercial electret microphone. That sparked an interest in the technology at AT&T. We had finally convinced people at Ma Bell that we could replace the 57-cent carbon button microphone with an electret microphone, but the folks in product development couldn't make the device work. We were immediately called in to help diagnose the problem. The only difference we could see was that the product version differed from ours in having a gold electrode rather than aluminum. We had used aluminum because it was easier to deposit on the electret and less expensive, but it also was more reactive with the environment so it could wear off and the device would stop working. That was the logic behind choosing gold for the commercial microphone. We were stumped as to why this seemingly modest change made the microphone cease to work.

We approached two Bell Labs material scientists for help resolving this issue: Frank Ryan and Harold Schonhorn. They showed us that the polarizing charge was able to migrate through the thin gold film essentially rendering the electret material unpolarized. The problem was solved by depositing a thin layer of titanium under the gold layer, the combination producing a good blocking electrode.

Our work on directional microphones required us to work in a space, an anechoic chamber, that was devoid of reflections so that we could control the direction from which sound emanated. At Bell Labs we had a large anechoic chamber. It had a metal mesh floor that bounced when you walked. Several feet below the floor were wedges of foam and fiber sound absorbing materials. The walls also had wedges on them, as shown in Fig. 7.2. We kept a conch shell in the chamber so that we could show visitors what happened when they walked into the chamber and closed the door. The sound of the sea normally audible by holding the conch to your ear would disappear!

While working in the anechoic chamber we noticed that we would very occasionally see some weird, low frequency sounds appearing between 10 and 50 Hz. Such low frequency sounds aren't unusual in buildings, as they often

Fig. 7.2 Testing in the Bell Labs anechoic chamber

emanate from air handling equipment, but we'd expect the noise to be constant rather than coming and going at very infrequent times. We chased down everything we could think of as a source of these sounds including earthquakes, traffic vibration, and nearby construction, but nothing could explain what we were seeing. Finally, we looked to infrequent sources outside our immediate area and suspected we had found the culprit. It was blastoffs of Saturn V rockets in what was then Cape Canaveral in Florida!

We set up microphones in the anechoic chamber to be timed with the next Saturn V blastoff to determine whether our suspicion was correct. We found that indeed, the sound we were seeing was present shortly after the takeoff, although the speed of the sound propagation was a bit off what we would normally expect. Further, we saw that the sound arrived more than once. It came directly from the blastoff site and then wrapped around the earth and came again. We worked on this for a bit and were ultimately able to explain the sound speed difference from the theoretical sound speed in air.

Gerhard and I decided we should present our findings at an ASA meeting, as it was curious and interesting, so we submitted an abstract for the conference. The ASA conferences invite potential speakers to submit an abstract of no more than 200 words. A committee then organizes these into sessions with similar talks grouped together. This means that talks can contain the latest scientific information since a written paper is not required. It has always been my favorite forum for communicating results, partly for this reason.

Shortly after submitting the abstract, I received a phone call from a CIA agent asking me to withdraw the abstract and not to give the talk. My immediate reaction was "Hell, no! This is science. Why should I withdraw it?" The agent responded by telling me that the CIA could force me to withdraw the abstract, but they'd prefer that I do it voluntarily. I pushed them for a reason and was told that they were using the same sort of sound we were measuring to monitor atomic bomb testing in the USSR, and they didn't want that country to know how they were doing the monitoring. In the end, we withdrew the talk and never did publish our results. I didn't want to be just another Black man sent to jail.

By the early 1970s, the work Gerhard and I were doing was beginning to be recognized outside of Bell Labs and rewarded. I received two recognitions that year. I was made a Senior Member of the IEEE Acoustics, Speech, and Signal Processing Group. Senior Member is the highest membership category. It requires a nomination and is reviewed by a committee that determines if the title should be given. I also received, with Gerhard, the Callinen Award of the Electrochemical Society, Dielectric Science and Technology Division. The Callinen award had only been established in 1967 and recognizes outstanding achievement in dielectric science and technology as presented in a paper.

One of the side effects of the recognition I was beginning to receive was an invitation by William Shockley to debate him. Shockley was a physicist and a eugenicist. He was the manager of a research group at Bell Labs that was formed in 1945 to find an alternative to the fragile, glass tube amplifiers that electronics relied upon. He and team members John Bardeen and Walter Brattain developed the theory for the first transistor, a semiconductor device used to amplify or switch electrical signals and demonstrated its use. Shockley, Bardeen, and Brattain were awarded the Nobel Prize in Physics in 1956 for their research on semiconductors and discovery of the transistor effect.

Shockley left Bell Labs in 1953 and subsequently held positions at Caltech, Fairchild Semiconductor, and Stanford University, all while continuing to have an office at Bell Labs he would occasionally use in the summer. In the 1970s and 1980s, Shockley became known for his extremist views on race and its connection to intelligence. He argued that the higher birth rates among inferior people (as defined by their race) was driving the genetic pool into progressively inferior status. As he spouted his theories about Black racial inferiority, Black researchers reacted by challenging Bell Labs management. I wrote directly to Shockley and told him what I thought of his theories. It was in response that he invited me to debate him publicly. I thought very seriously about this proposal, but realized that if I did well, Shockley would likely attribute it to my being fairly light-skinned and thus having some superior white genetic origins. Instead, I enlisted a dark-skinned, female psychologist from Howard University to debate Shockley. She did some research on Shockley and learned that he was selling his sperm, but that his offspring were not considered to be unusually intelligent, thus disproving his genetic theories. It was a lovely moment in the debate when she presented that evidence.

Despite encountering persistent racism, the late 1960s to early 1970s were a lovely time in my life. Gerhard, Bernard and I were making good progress on understanding the physics behind electret behavior and were beginning to be recognized for our accomplishments. My family was thriving, and life was exciting.

8

Diversity Advocacy at Bell Labs

In 1969, AT&T became one of the first large companies to establish employee resource groups—voluntary, employee-led groups to connect employees with similar interests. Its first such group was for Black employees and is now called Network Black Integrated Communications Professionals (NETwork BICP). The stated mission of this 55-year-old organization is "Transforming our future by empowering our members and our community today."

Also in 1969, Earl Shaw graduated from the University of California at Berkeley with a PhD in physics and joined Bell Labs research. Earl was a native of Mississippi and the son of sharecroppers. His education started at a three-room schoolhouse on the Hopson Plantation but improved as he and his mother moved to Chicago. He was able to go to the Crane Technical High School and then to the University of Illinois, after which he received a master's degree in physics from Dartmouth before his time pursuing a PhD at UC Berkeley.

Earl's time at Berkeley overlapped with the Berkeley Free Speech Movement. The movement arose as a protest of university restrictions on the political activities of students on campus. In 1964, sit-ins and demonstrations led ultimately to large rallies that forced the university to overturn policies that would restrict the content of speech and advocacy, and to commit to free speech as a fundamental value. It was as a witness to the power of people to affect change that Earl arrived at Bell Labs.

When Earl arrived at Murray Hill, there were only around 20 Black people among the technical staff at Bells Labs. There were many more Black people to be seen among the clerical and administrative staff, often in very low-level or dead-end jobs. Clearly, Bell Labs was not representative of the community

J. E. West, I. J. Busch-Vishniac, *Mic Drop*, https://doi.org/10.1007/978-3-032-20922-1_8

in which it existed or even the nation at large, and Earl wondered if it might be possible to call attention to this and effect change by bringing Black employees together. He invited Black employees at all levels to come to his home for discussion and socializing. It was from this that the Association of Black Lab Employees (ABLE) was born. It is unclear whether any of the attendees at the first meeting were aware of the employee resource group formed for Black employees at AT&T, and the two groups neither competed nor worked together to any significant extent.

The first few meetings of ABLE centered around defining common views. Fairly quickly the group agreed to advocate on behalf of those already employed by Bell Labs and to seek employment of a greater number of Black people at all levels, especially on the technical staff and in management, where their numbers were so small. People attending the gatherings and emerging as leaders included Link Hawkins, Charles Craig, Tim Pernell, Dick Dennis, and Ray Story, besides Earl and me. Of these, at least a couple were from outside the technical staff and could bring in the views and issues of administrative and clerical workers.

As ABLE came out of the shadows and Bell Labs managers became aware of its existence, we grew careful. We never had an identified leader, which puzzled Bell Labs management. Each meeting would be run by a different person, and we would discuss who we wanted to participate in various meetings or activities. We wanted to remain an amorphous group with unified goals and ideals. We also understood that management liked to keep tabs on organizations by having spies who would report to them. ABLE knew that there were some such spies in the group of regulars, so we set about to confuse management with some of our proposals. For instance, we knew that as a monopoly, the laws stipulated that AT&T could not accept overpayment of a bill. We would suggest that an option, should progress in our goals not be forthcoming, would be for ABLE to encourage Black people to overpay their phone bill by ten cents. The system would end up spending well more than ten cents per person returning that overpayment. We never intended to act on this proposal, but we left it as a sign to be delivered to management that we had options that would make their lives uncomfortable. We were also careful to protect the junior members of technical staff as much as possible. The theory was that they should be given the chance to excel in the job for which they were hired before we let them take on some of the advocacy activities in which we were engaged.

My involvement with ABLE was initially difficult. Colorism bias reared its ugly head, and I was viewed with suspicion. I was frequently left out of planning and considered likely to be one of those direct conduits to the management—i.e., a house nigger. I refused to accept this rebuff and basically said I was going to do what I thought was right regardless. Eventually people came around and saw that I was making progress, although I was often said to be making progress only because I was part of the system. This was neither my first nor my last dealing with colorism. It's like a bucket of crabs fighting each other and not quite understanding that the real enemy to them all was the bucket owner.

I was also criticized for my focus on Black people gaining entrance to the highest levels of the lab hierarchy, the research scientists and science managers, but I felt the criticism was illogical. I was told that it didn't make sense to focus on the technical people at the Labs because there were so few Black people there. I replied that that was precisely my point. We needed to have more people at that level.

President Bill Baker was supportive of my activities advocating on behalf of Black employees of Bell Labs. Interestingly, the middle managers I had had, including Max Matthews and Manfred Schroder, suggested that I spend more time on science and less on my crusade. This seemed to be the pattern. The top management were much more receptive to my diversity work than the rank and file. Within my home department and center, there was a strict separation of research discussion from social commentary. I was clearly valued among colleagues for my research, but not so much for my social views.

We petitioned the Bell Labs Council for a meeting to present two concerns. The Bell Labs Council was made up of the most senior executives, including President Bill Baker. Our first issue was the status of Link Hawkins. At the time, Link was a supervisor, the lowest level of management. We pointed out that Bill Baker came to the company at the same time as Link, and while Bill's research was very good, it was not nearly as important to the company as Link's was. Yet Bill was now the Bell Labs President and Link Hawkins was in a modest managerial role. We asked how this could be explained. I thought at first that Bill was not paying attention because he had his head deep in a paper, but when he heard our request, he perked up. Bill said, "I know Link Hawkins very well. I consider him one of my dear, dear friends. And I have always been told that he is very happy where he is."

It took me nearly a month to work up the courage to speak with Link about why he was not interested in being promoted to a higher managerial position. He responded by saying "If they whispered 'promotion for Link Hawkins' ten doors down from me, I would have grabbed it immediately." This made it clear that someone in the management chain was making erroneous assumptions or trying to deny Link a promotion. I conveyed the information up the chain of command and a short time later, Link was promoted to Director, a significant bump up the management ladder.

Our second request was hiring of more under-represented minority technical people. To this, the Council replied that "We hired you. Find qualified people and we will consider hiring them." This issue turned out to be difficult, for indeed, it was nearly impossible to find qualified Black men and women for the technical staff. In the 1970s, the number of PhDs offered to people of color in the sciences accounted for fewer than 2% of the total number of such degrees awarded to US citizens and permanent residents. We determined that we would need to grow these people through programs we would craft.

Our first attempt at such a program was to suggest that AT&T Bell Labs establish a center of excellence at an HBCU. This would afford a lot of interaction with students which we could use to encourage them to complete degrees and join Bell Labs. We chose to work with Morehouse University, an HBCU in Georgia, but we were unable to complete an agreement because the overhead demanded by the school would siphon about half of the funds away. We approached Howard University with similar result. Clearly, this approach was not going to be acceptable.

Finally in 1972 with Bell Labs approval and funding, we created the Corporate Research Fellowship Program (CRFP). CRFP was designed to support Black students in PhD programs in science areas relevant to Bell Labs. The aim was to award fellowships to ten to twelve new students per year and to provide full support to those students for the five and a half to six years required for degree completion, including funds for tuition, fees, books, conference attendance, summer employment, and an annual stipend. Further, each student would be assigned a mentor with the objective of ensuring that Fellows had a substantive relationship with an experienced scientist in a related discipline, who could provide guidance, nurturing, inspiration, and advocacy. Mentorship would be reinforced by bringing new CRFP Fellows to Bell Labs to work with their mentor during the summer after graduating with the undergraduate degree and before attending graduate school. Additionally, students would have an academic advisor from Bell Labs who could serve as a neutral party advocating for students on behalf of the company for issues such as changing research direction, changing graduate advisor, or making curricular adjustments to suit personal needs.

CRFP did not start off well. I was invited to a dinner to meet the students in the program and arrived expecting to meet ten or so students but found instead only three. I was told by the director of the program that they had scoured the country but could only find these three students. When I asked where they had looked, the response was "Harvard, Yale, MIT, Stanford, and similar." The problem was immediately obvious to me. These were the schools the white technical staff had come from but would not be where we would find Black qualified students. Those students would overwhelmingly come from HBCUs including Morgan, Talladega, and Howard. Recruitment was revamped and CRFP numbers ramped up quickly in the following years. There were certainly those who believed that we were lowering the standards by recruiting at HBCUs rather than just the Ivies, but the accomplishments of the Fellows over the years have proved that thinking to be wrong.

An additional problem CRFP encountered was pushback from an organization similar to ABLE but focused on women in science at Bell Labs. We originally approached them with the idea of working jointly on programs to enhance diversity at the lab, but they thought our ideas so wild that there was no chance Bell Labs management would support the programs, so they declined the offer. Once we had received very significant funding from management, the leaders of the women in science group demanded to be part of the CRFP program. At that point, ABLE insisted that they create their own program so that the necessary differences in recruitment and the need for significant numbers would not be compromised. It was a stance that did not earn ABLE friends among the women scientists at Bell Labs, but it led to the creation of the Graduate Research Program for Women (GRPW), which paralleled CRFP.

As we were ramping up CRFP, ABLE also determined that we needed to prime the pipeline by capturing students still in their undergraduate years and showing them the lure of research to encourage them to attend graduate school. We crafted the Summer Research Program (SRP), which was opened in 1974. In this program, for which I wrote the proposal, minority and female students who had completed 3 years of education in mathematics, science, or engineering would qualify to spend ten summer weeks working at a Bell Labs site. They would receive housing (arranged at Rutgers), transportation, and a stipend. Each student would receive a research topic and be working with a Bell Labs scientist as a mentor. The program would end with presentations by the students and social events would be scattered throughout. In the early days, the aim was to bring in about 60 students per summer for this program.

In addition to these critical programs, ABLE advocated for educational programs for staff, with a focus on Black support staff. Educational programs were broadened, and we encouraged people to take advantage of them. In several cases, employees who started at dead-end or low-level jobs were able to shift their employment after participating in these programs. For instance, Tim Pernell started at Bell as a groundskeeper, mowing the lawn. He took advantage of the educational programs offered at Bell Labs and ultimately retired as a Member, Technical Staff. More common was the elevation of Black secretaries to managerial positions. Among the things we were able to do is teach staff how to use the welfare system for additional support during their educational programs. This meant that employees could potentially quit to change from part-time to full-time degree programs, after which they could possibly return to Bell Labs in a much higher level and better paying job.

CRFP and SRP were rousing successes that lasted for many years and that not only produced Black scientists for the nation, but that led very specifically to Black scientists who joined Bell Labs, including Bill Wilson and Bill Massey. Three factors were responsible for the success of these programs: commitment to them at the top levels of management, appropriate funding, and a focus on creating scientific communities. For example, CRFP's commitment to active mentoring was viewed as a key ingredient of the program, but it was initially hard to find mentors for all the fellows. I was among the group that went to see Arno Penzias, then President of Bell Labs and a Nobel Prize winner in Physics, to discuss this issue. Arno said that he wished all problems were so easy to solve. He added a question to the annual evaluation of every technical employee asking what they had done in support of diversity goals during the year. With that single change, we found that we had more scientists wanting to be mentors than we had students.

As ABLE created the CRFP program, we were aware of the racist assumption that Black students from HBCU schools would perform worse than white students even when given equal opportunity. Certainly, it was not the only time in my life that the expectations for me and for Black people in general were set low, often by well-meaning people creating programming. The argument among many was that the inferior educational training afforded to Black students would hamper those scholars forever. This sort of argument leads to the expectations for Black students being lower than for white students. I've fought this racism my entire life. To the extent there are differences in the performance of Black and white students in school, I know these result from differences in educational opportunities. Creating different expectations just feeds into the idea that Black students can't perform as well as white students and it encourages talented Black students to assume they can't do well. It becomes a self-fulfilling prophecy. In CRFP, we fought against this assumption by helping fellows from HBCUs to see where their education had left gaps and to urge them to take a series of high-level courses to plug these holes before starting their graduate student experience. Our approach worked well. Almost all the fellows with HBCU undergraduate degrees thrived in the program.

Our success in creating the CRFP and SRP programs was a turning point for me. It showed me that others were seeing what I was seeing. I was not alone in thinking that Black people could be just as accomplished as white people if simply given the opportunity. It showed me that at Bell Labs, we were heard. More than that, we were supported by the Bell Labs leadership.

Bill Massey has been quoted as saying, about the CRFP program, "We weren't just trying to turn one Black student into a scientist – we were trying to create a Black scientific community." This distinction was important. The aim wasn't to find a person, push them forward, and then release them from the club when done. The aim was to build the infrastructure needed to support Black scientists from the ground up and to have that infrastructure stay in place by ensuring that mentors, students, alumni of the programs, and potential participants in the program all felt themselves to be part of an ongoing and vital team. It was this sense of community that enshrined the CRFP and SRP programs and helped make newcomers feel welcomed.

The SRP and CRFP programs grew and thrived for decades. SRP spread throughout Bell Labs and was very popular among AT&T leadership, students, and universities. Trivestiture of AT&T in 1996 led to the establishment of two CRFP programs—one for AT&T and one for Lucent's Bell Labs. The AT&T program combined the graduate fellowship programs for minority students and for women into a single program called the AT&T Labs Fellow Program (ALFP). Lucent continued two separate programs. Ten years later, all of the programs were gone.

The CRFP and SRP programs produced tangible results. CRFP supported more than 260 graduate students of color toward MS and PhD degrees. More than 10,000 students went through SRP. By 1992, 10% of the PhDs awarded to racial minorities in engineering in the US were for CRFP fellows. The programs stood as models for others that followed, including the Meyerhoff Scholars program at University of Maryland, Baltimore County.

Over the years, I've been stopped by people who want to thank me for helping create and support the SRP and CRFP programs. I've received letters and email thanking me and even been acknowledged in publications. Even though the programs are long gone, I still get the occasional thank you.

Although the CRFP and SRP programs came to an end, the community formed by ABLE founders, former fellows, their students, and their colleagues has persisted. There are now generations of students and scientists that have been connected by the programs. To give a few personal examples: I have taken on students from Morgan State University who were trained by people I knew or mentored when they were CRFP Fellows. A granddaughter of mine had a former CRFP Fellow as her advisor at the University of Illinois. At the

Festschrift for my 80th birthday, a young man from Morgan State had a poster presentation of his research. He was shepherded through the graduate school process by CRFP alumni he met at the celebration and is now a faculty member bringing his own students along. These sorts of stories are commonplace and speak to the persistence of the Black scientific community we created.

There are other indicators of the success of our team in creating a community of Black scientists as well. For instance, of the 30 or so Black inductees into the National Inventors Hall of Fame, four or five of them came from Bell Labs so over 10% came from a single company.

ABLE didn't simply disband as programs were created and management of them established. We remained a vibrant group and tackled other items as seemed appropriate. For instance, ABLE played a role in the establishment of the Black United Fund of New Jersey. In the early 1970s, United Way was nearly the only philanthropic organization that companies such as AT&T would sanction to the point of matching contributions of employees and allowing for direct withdrawal from employee paychecks. But the United Way organization did not seem to be paying much attention to the needs of the Black communities, so a new national organization was founded in 1972 called the National Black United Fund. Its mission was to foster contributions to Black organizations engaged in social change and the development of Black human potential. In the 1980s, Charles Craig, Earl Shaw and I, partly through ABLE, successfully pushed for the establishment of a branch of the Black United Fund at AT&T as an alternative to United Way. Bob Pickett, a local judge in the area (the one who intervened in musician George Clinton's arrest for cocaine possession) was also instrumental in establishing that Black United Fund. We were successful and participation and geographic impact grew over the years. The Black United Fund of New Jersey is still quite active and plays a significant role in the National Black United Fund.

I have always been proud of my involvement with ABLE and the creation of the CRFP and SRP programs, but I hadn't thought about its impact broadly in science until recently. Bill Wilson organized a conference called Quantum Noir and asked me to speak. I was asked what I considered my legacy and mentioned my work on electret devices. The response was to point out that I had probably impacted more people through the creation of CRFP and SRP, and I believe that is correct. It is great at my age to be given something additional to consider my legacy.

The politics are changing in the US and there has grown a strong backlash against diversity initiatives. I believe that programs such as CRFP and SRP have established enough of a community that the gains we have made will continue even without additional structure. It will slow down diversity progress, but it can no longer stop it. Borrowing from gay people, I say "We're here. We're Black. Get over it."

9

Divorce

The mid to late 1970s were a time of great changes in my life, both professionally and personally. In 1975, Gerhard and his family opted to return to Germany, where Gerhard assumed an academic appointment at the Technische Hochschule in Darmstadt. This didn't dissolve our collaboration, but it did mean that I did a lot more extensive traveling. For a decade I tried to spend most of my summers in Germany working with Gerhard and his students. Bell Labs was fine with my travel, but my family suffered.

Louisa's grandmother passed away before her grandfather, who died in 1965. By the mid-1970s Louisa's immediate family migrated north from North Carolina to New Jersey. Louisa's mom, Mabel, moved into a place in Roselle, NJ, no more than ten miles from where we were living. She taught math at a local high school for boys and lent a hand as needed with our household. Louisa's brother, Harold, and his wife Barbara, also looked to move to New Jersey and lived with us for a while.

It was a tumultuous time in the US. There were protests of the war in Vietnam and a very active group of young people who were preaching the need to fight the government. Drug use became more mainstream, with a focus on marijuana, LSD, hallucinogens, and cocaine. My kids, like pretty much all teens at that time, were exposed to the drug culture. I adopted the attitude that if they wanted to experiment with drugs, I wanted them to tell me and let me get the drugs for them. I did this knowing that a distant relative had had someone slip drugs into her drink without permission, an act that ruined her life. This way I could protect the kids from the unscrupulous drug dealers who would often change up what they were selling you with no warning. The kids did try marijuana and at least once, peyote, but they made it

J. E. West, I. J. Busch-Vishniac, *Mic Drop*, https://doi.org/10.1007/978-3-032-20922-1_9

through this period of their lives unscathed. Judging from the damage done to neighborhood kids by the drug culture, I think I did the right thing.

After years of fighting in Vietnam, Saigon fell in 1975, effectively ending the war in defeat for the US-supported side. Politics had gotten quite bitter, and President Nixon had been forced to resign from office in the summer of 1974, with Gerald Ford taking over. The US economy had weathered a recession from 1973–1975 and then produced stagflation—high inflation with uneven economic growth. An oil embargo created lines of people waiting to fill their cars with gas, adding to the malaise. Student unrest had led to a tightening of funding for public and private universities at the same time the demand for seats at the institutions was rising rapidly. To deal with the financial pressures, universities sought to cut all but the "essential" activities and personnel. Louisa, who loved her counseling job and was beloved by the students, was let go by Rutgers University. Already prone to depression, this unanticipated change in her status sent her reeling.

Shortly after Louisa lost her job, and while her brother and sister-in-law were living with us, we had a terrible flood in Plainfield. I was so wedded to my work and so anxious to escape the tension of my souring relationship with Louisa that I tried to wade through the high waters to the place at the top of the hill where I had left the car so that I could drive to the office. I didn't get very far and was forced to return home. By then water was trickling into the basement and we were hastily moving Harold's and Barbara's stored belongings upstairs. We heard a loud crash as the house's foundation gave way, letting water flood the basement faster. It was a horrible mess, but in a moment of wonderful levity, Mabel somehow managed to get through the flood and to our house. She came in explaining that she had just found some adorable jelly glasses and she bought them for us. Mabel is shown in Fig. 9.1 along with my daughter Melanie and my mother Matilda West.

While this was not the last time we saw Mabel alive, it was close to her very much unexpected demise. In 1976, Mabel was stabbed to death in her bed by a young man she had hired from her neighborhood to help her get odd jobs done. We were, of course, shocked and grieving at this turn of events. It was difficult not to focus on the murderer and wish for him to suffer, but Louisa correctly pointed out to the kids that he had been a foster child who had apparently been locked up in chains in the basement and beaten. Nobody could anticipate that such treatment would yield anything but a seriously mentally ill person.

Mabel's death coupled with Louisa's job loss and our failing marriage plunged Louisa into a great depression. I suspected that she had been self-medicating for some time, but now I scoured the house for any drugs she

Fig. 9.1 Matilda West, Melanie, and Mabel Moss

might have hidden. I found about 200 pills around the house. I got rid of all of it and tried to help Louisa break her addiction, but she simply moved from drug addiction to addiction to alcohol. This was not my only involvement with drug and alcohol addiction, having grown up with an alcoholic father, but it was the first time I tried to help someone to escape addiction's clutches. I did so while using drugs myself, but in a very controlled fashion. At Bell Labs, there was a culture that pressured people to produce results and that rewarded working long hours. I learned that many of my colleagues were using cocaine to improve their performance, as it enhances focus and energy. Certainly, cocaine was readily available within the walls of the lab, and I admit I occasionally took advantage of it.

By the late 1970s, my household's dynamic had changed. Melanie had gone off to Rutgers, where she crafted a special program combining music and engineering. It was not easy to get approval for an unorthodox degree program, and I remain very proud of Melanie for having had the persistence and ingenuity to pursue her dreams. Laurie and Jay were still in school but both were now essentially teenagers with the attitudes that accompanies that.

During this time, the fights between Louisa and me had grown intense and sometimes were even physical. When I wasn't present, Louisa would find other ways to make it clear she was mad at me. I was told by the girls, for instance, that they heard hammering one morning and went downstairs to see what was going on. They found Louisa using a hammer to smash every piece from a marble chess set she had given me as a gift. When asked what was going on, she just told the kids that "Everything is fine. Go back to bed," but the hammering kept going.

I had started coping with the situation by avoiding time at home, hanging out at some of the bars in the neighborhood. Unfortunately, this meant the kids saw even less of me than normal. I had sought and found companionship elsewhere and wanted to move on. My infidelity didn't help the situation. It made Louisa feel rejected and added to her pain and anger. I had tried to keep my dalliances discreet but had apparently failed. It was a lesson I had learned growing up. In my family, men were expected to always put family first and to support them, but once obligations were met, they were free to do whatever they wanted as long as they kept it concealed. It is how I've operated throughout my life.

Ultimately, I told Louisa that I wanted a divorce and that I was leaving. I moved in with Kirstie Gentleman, who owned a house in Westfield, NJ. I had met Kirstie when she had come to interview me as part of her work in communications at Bell Labs. Kirstie's job was to write public interest pieces. We had, at that point, been seeing one another for a while and were quite compatible.

All told, Kirstie and I lived together for 3 years. We moved from Westfield to Hunterdon County, where Kirstie bought a house. I had the kids part time, and they tried to get along with Kirstie, succeeding only modestly. Laurie now admits that she was bitchy to Kirstie, resenting her for having an affair with her father, a married man, and driving her mother into depression. Laurie found the normal issues of being a teenager and simultaneously coping with her parents' divorce and mother's depression difficult. She was not coping well with the home environment.

As Louisa's infirmity became more and more acute, I sought to change the living situation so that I had primary custodianship for the kids. It took a long time for Louisa to recover from the various traumas in her life. My youngest at the time, Jay, felt abandoned. He was struggling in school and when asked why he wasn't trying to do well, he explained that nobody cared. Laurie and he were doing their best to survive, but parenting in Louisa's household was noticeably absent. Laurie was trying to support Jay, given that Louisa was often unavailable even when present, and the pressure of being a teenager and caregiver was simply too much. Melanie was in college and thus protected some from the upheaval, and Laurie was to leave in a couple years for college, but Jay was younger and more vulnerable.

Given Laurie's concerns, I had her move in with Kirstie and me. That meant that Laurie attended high school in Hunterdon County, where she was almost the only Black person in her class. The schools were not terribly good, but Laurie managed to weather them and graduate near the top of her class. She got particularly interested in astronomy while taking a class on the subject and we would go out to parks late at night to get star trails with a telescope and camera. I had always enjoyed photography and found this a wonderful way for the two of us to share a common joy.

I was advised by my attorney that it would help if I had a house of my own, so I purchased a house in Plainfield and I built a rudimentary sound stage in the basement for the kids. The house also had an in-ground pool and a garden. My relationship with Kirstie had run its course by that time. She was under constant pressure for living with a Black man, and I found my friends changing their attitude toward me, nearly ostracizing me, as I was living with a white woman. I felt as if I were being asked to choose between personal values and community values. In the end, Kirstie and I parted amicably. The courts made an unusual decision for the times and awarded primary custody of the kids to me, the father, rather than their mother.

In addition to my kids, the Plainfield household ended up for a time with an additional teenager residing there. Laurie came home one day with her friend Gina Gupton and asked if Gina, who was having serious issues at home, could move in with us. Remembering the kindness paid to me by my aunt when I needed a room in Philadelphia so I could attend college, I agreed if details could be worked out with Gina's family. That's how Gina, who I think of as my third daughter (chronologically), came into my life. We have remained close since she stayed with us, and I have been thrilled to watch her career blossom as an Emmy-award winning producer and director. Gina has an extensive background in television, film, and live concert production. One of Gina's unique characteristics is that she works inordinately hard to ensure that the people she loves have everything they might desire.

A little later, after Laurie had left for college, Laurie had a serious car accident that left her with a broken leg, broken ankle, and stitches on her face and leg. She was near the end of a term, but she neglected to tell the school of her accident and was recorded as having failed because she stopped coming to classes and taking exams. Laurie grew depressed and I worried about her. Fortunately, she overcame the depression in time and got back to being herself, although she didn't return to school.

The sound studio in the house turned out to be a wonderful addition. Laurie and Jay spent much time in the studio. It also attracted local talent, including two people worthy of mention: Bernie Worrell and George Clinton. Bernie Worrell grew up in Plainfield. He gave a classical piano recital as a young man and was declared a musical genius. He is credited with being a founding member of the Parliament-Funkadelic Collective. A keyboardist and synthesizer master, he had a significant influence on both funk and hip-hop.

I met Bernie through the church Louisa and the kids attended. Early on, he served as the accompanist for the choir Louisa led. Sometimes, instead of just dropping Louisa off, I would sit in the church and listen to the music being generated. Interestingly, Bernie was seen as disruptive. He caught on to musical pieces very fast and seemed not to understand that others weren't working at his speed. Most people who ran into Bernie couldn't understand what he was talking about. He gave the appearance of being very flippant and unable to focus, but when you really listened carefully to what Bernie was saying, he was extremely focused and on point. I considered that a sign of brilliance, a sign of somebody who was far above the normal person in intellect, and I began to identify with that in a very positive way. I thought, Bernie's considered a genius. If you look at my thoughts and writing, I think like he does. I wonder, am I a genius?

Bernie spent time teaching Laurie and Jay in our home studio. He interacted well with my kids although his son is somewhat younger than my kids. They got sufficiently close that Laurie and Jay formed a band called Room Service, with Laurie singing and Jay on the keyboard. Bernie would come by and give them tips and pointers. Frequently Jay and Laurie would perform several songs in the middle of the set of Bernie Worrell and Friends at clubs in New York City.

George Clinton was born in North Carolina but grew up in Plainfield. He is a singer, songwriter, producer, and bandleader. Clinton was a major influence on hip-hop and P-funk music through his band known as Parliament-Funkadelics. Much of what he used in the Funkadelics was created by Bernie Worrell. Clinton was inducted into the Rock and Roll Hall of Fame and awarded a Grammy Lifetime Achievement Award. Although he wasn't personally seen in our home studio, some of his band members would show up.

The downside of this exposure to the music scene was the simultaneous exposure to the drugs that were prevalent among the musicians. I recall that one time, a musician in the Parliament-Funkadelic band shot up heroin while seated on a couch in my house. Another time, I was involved in bailing Bernie Worrell and Mike Hampton, the lead guitarist, out of jail. These were not the examples I wanted the kids to follow.

At some point in this time period, Laurie and I had a serious falling out over a telephone bill. We screamed at one another and Laurie stormed off to live with her mother instead of with me. Communication between us was nonexistent for a long time, but we did finally bridge that gap because even in our worst moments, we knew we loved one another and cared about each other.

An additional source of tension during this time was my mother's situation. I can't recall how I came to learn of her involvement with Reverend Jim Jones, but I did come to hear of it. Mom had invested in the People's Temple, which we now know was a cult, and was committed to their doctrines. She had grown up in a religious family and always been grounded in a church, so her attachment to the People's Temple wasn't totally outlandish. It took a lot of effort and many visits, but I convinced Mom that this was not a helpful or good group of people. I managed to get her out of their clutches before the Jonestown Massacre of 1978, in which nearly a thousand followers of Jim Jones died in a mass suicide/murder event in Guyana. I still get chills thinking about how close she might have been to being among those poor souls who died.

Overall, in the late 1970s my home life was marked by chaos and transitions. I moved twice, my mother-in-law was murdered, two of my children left home, my marriage ended, and I rescued my mother from a cult.

10

Directional Sensors

In addition to the summers and odd conferences Gerhard and I spent together, I brought two of Gerhard's students to Bell Labs for internships: Raimond Gerhard-Multhaupt and Heinz Von Seggern. Raimond stayed for something like 6 months. Heinz and I worked together for a longer period of time.

By the mid-1970s, Gerhard and I were beginning to have a better understanding of the physical processes involved in electret behavior. Gerhard and I had determined that there are two prerequisites for useful electret behavior: dipole moments and deep traps.

In physics, we often ignore the dimension of an object and treat it as though it exists at a tiny point in space. When speaking of molecules, this often makes sense as a typical molecule is on the order of a billionth of a meter in size. If the molecule is ionized, and thus exhibits a net charge, we tend to represent that as a charge at a point. For some materials, their chemical structure means that the charges they exhibit are very slightly displaced from one another. These molecules are said to have a dipole moment. An example of a material with a dipole moment is sodium chloride, table salt. Materials with a significant dipole moment are very common and include virtually any crystalline material.

For materials with a dipole moment, even those with net zero charge, it isn't useful to think of them as existing at a point. Electric fields exist around any charged material or dynamically changing magnetic field, and they are pervasive in our environment. For instance, there is an electric field on the surface of the earth simply due to the ionosphere interactions. When materials with a dipole moment are subjected to an electric field, they try to orient so that the force on them is minimized. This means they align with the field

J. E. West, I. J. Busch-Vishniac, *Mic Drop*, https://doi.org/10.1007/978-3-032-20922-1_10

to the extent that they can. Bernard, Gerhard, and I figured out that to make useful electret materials, we needed materials with dipole moments we could align by exposing them to intense electric fields. This meant we were looking for materials we could polarize easily. Importantly, the alignment had to be retained once the applied field was turned off, so materials with a dipole moment that would relax after alignment were not useful to us.

The second thing we found was an essential part of useful electret materials was the existence of deep traps for electrons. Some materials, especially semiconductors, can trap ("grab") electrons to which they are exposed. The ability to retain these electrons depends on how much energy it takes to remove them. If the energy is slight, the traps are said to be shallow. If the energy required for removal is high, the traps are said to be deep. Our research team determined that permanently polarized electrets were obtained in materials which had deep traps. The process of polarizing the material requires opening the traps, which one of Gerhard's students showed required heating the material, then jamming in the electrons, and then closing those traps and cooling the material to leave the material permanently charged. The materials we were using, forms of Teflon, had deep traps. We found that wet-formed silica could also work, but it had disadvantages from a manufacturing standpoint.

Heinz and I worked on positive charge traps as well. These are slightly different as positive ions tend to have shallower traps than electrons, but we figured out how to accomplish positive electret material charging and published a series of papers.

We modified an electron microscope so we could very precisely control the flow of electrons to the films and foils we wanted to study. We developed measurement methods to determine how deep the traps were in the materials, typically using thermally stimulated currents, a process in which we looked at the charge released, the current, as a function of temperature rise. Because the electron microscope requires a vacuum, I set my workday to begin around noon. That meant that lab assistants could turn on the microscope and have it establishing a vacuum before I arrived. Of course, this meant I often didn't get home much before midnight.

On occasion I would bring the kids with me to the office on a weekend or school holiday. The lab was across the hall from my office. I would allow the kids into the lab while I was loading the electron microscope with samples, but I would insist that they be in my office when the microscope was fired. My concern was the possible exposure to leaking radiation, although we regularly tested to be sure that this was not a problem. When asked why they couldn't stay in the lab to watch, I told them the system could blow up and I wanted to keep them safe. The unexpected result of this was a drawing Jay made for one of his class projects showing me in the lab with things blowing up around me. I framed that picture and displayed it for years.

It was during this time that the Hungarian Academy of Sciences invited me to a conference in Budapest. The invitation came with a wax seal and was very formal. I had been planning to visit Bernard in Vienna, so the Academy sent me a plane ticket from Vienna to Budapest, but I had read Agatha Christie and opted instead to hop on the Orient Express, a train that would connect the two cities.

The day I was to travel was very hot, so I wore cutoff jeans and a fishnet shirt. The trip went well until I hit the border between Austria and Hungary. I was pulled off the train and everything I was carrying was taken from me. I was not alone at being removed, but was one of about 30 people treated this way, although most of the others were circus people. They split the group in half and my half was put in a small concrete room where they went through all my belongings and asked lots of questions—where are you going? who are you seeing? I showed them my official invitation, but it was obviously thought to be fake. Then they cut that group in half, and then half again and again and eventually they put me on a little train to an unknown place. I rode through the night and as I awoke, I saw the skyline of Vienna. I had been sent back to my starting point.

I didn't know it as I arrived, but my things were also on the train, including my only decent pair of pants and clean shirt. I used the bathroom in the station to clean up and dress well and went to the Hungarian Embassy. It was still early, and the embassy was not open, but I pounded on the door until someone answered it. I told them what had happened, and they escorted me to the airport, put me on a plane, and got me to Budapest in time for my scheduled talk.

Although we still had plenty of questions about the fundamental physics involved in electret materials, word came down in 1981 that we were to consider that line of investigation now closed so we could focus more on research of direct interest to AT&T. This was the first time that boundaries had been put on my work, and I was not happy. However, I still had plenty of interesting research projects.

Gerhard and I worked hard on a nice application of electret devices—the development of gradient microphones. Compared to earlier telephone microphones, the electret microphone was more linear, meaning it had less distortion. It was also inherently more sensitive and could be made smaller. As microphones get smaller, they look more and more like point receivers of sound, meaning they would be equally sensitive to sound arriving from any direction. In many circumstances, this is not desired as it means that noise from equipment or an audience is picked up as well as the desired signal. One way to counter this is to make the microphone directional, in other words, to make sure it responds to sound from preferred directions. There are two ways to accomplish this—either by making the device itself directional, or by combining sensors in arrays and summing their signals appropriately. Gerhard and I worked on making devices inherently directional rather than using multiple sensors wired together. We figured that this approach was better because it would cost less since it used fewer microphones and would take up less space than an array. The space issue was particularly important for computers, where space for circuit boards and batteries and disks and any other equipment is at a premium.

Regardless of the approach, the fundamental idea of directional microphones is the same. In the simplest, first order gradient microphone, the idea is to have essentially two microphones collocated with the signal formed by differencing the output of the two. Imagine, for instance, two microphones placed back-to-back and stacked vertically. Sound coming from the side would travel exactly the same distance to each microphone, resulting in identical signals from each, which means the difference would be zero. In theory, this setup would not be sensitive to that sound then, given the signals were subtracted. Rather than two separate microphones, we realized you could place the electret film, coated with metal on both sides between two metal screens and produce the same effect. Sound coming from the side would apply the same sound pressure to both the top and bottom surfaces of the foil, which would thus not move and not produce a signal.

Gerhard and I produced gradient microphones of second order (e.g. differencing two first order gradient microphones) and third order (e.g. differencing two second order gradient microphones) in the lab by building microphones with multiple paths for the sound to travel to the sensing surface. By controlling the length of the acoustic paths in the microphone we could make it seem as though there were an array of microphones with spacing determined by the acoustic path length differences.

During this period, I also had a great opportunity to work with my brother on a project. I was approached by a doctor named Connor Cough and asked whether there was any information on whether polarized materials would promote healing. We agreed to do a very simple study. We took two petri dishes and loaded them with cells. We put polarized Teflon in one of the dishes and neutral Teflon in the other. Then we simply measured the growth of cells and compared the two. The result was very clear. The petri dish with the polarized Teflon grew orders of magnitude more new cells than the dish with the neutral Teflon. It seemed, in this simple experiment, that there might be reason to believe that polarized material could aid in cell growth, and thus in healing of wounds.

By this time, Nate was an endodontist on the faculty of Howard University. Our goal was to determine whether the presence of a polarized material would have an impact on wound healing. Nate and his students and I experimented in dogs by including polarized electrets in the packing after tooth removal. He showed this led to faster healing with less likelihood of infection. I don't know whether these results were ever taken to the next level of a clinical trial on humans.

In late summer of 1980, Ilene Busch-Vishniac joined Bell Labs as a postdoctoral fellow in the Acoustics Research Department. Ilene graduated from MIT with a PhD in mechanical engineering and had worked in noise control. She was offered a position at 3 units of Bell Labs and opted for the postdoc with us even though it was a temporary appointment, and the other two offers were for much higher-paying, permanent positions. Ilene started by finding out what everyone was doing so she could decide who she wanted to work with. I was in Germany when Ilene arrived, so I was the last person for her to interview. When I arrived back at Bell Labs, she asked to meet with me and I put her off, explaining that I had to first catch up on months of mail. Ilene returned day after day, and I kept delaying meeting with her. Finally, I arrived 1 day and walked into my office to find Ilene sitting there. She said I should just go about my business, but she wasn't leaving until I spoke with her. That was the beginning of a wonderful collaboration and friendship.

I talked with Ilene about my work on electret materials and devices. We spoke about directive sensors. Before we were done, Ilene had determined that I was the right person for her to work with. I taught her how to polarize electret materials and we came up with a great project: building a single electret device that mimics the behavior of a linear array of microphones.

At this point in time, conference telephony was just coming of age. The idea was to establish audio connection of people in one room with people in another room so that large groups could meet without leaving their home offices. The problem acoustically was an interesting challenge. A microphone recording sound in the speaker's room would record sound that had been affected by the room, i.e. by reverberating off the walls, ceiling, and floor. It would then reverberate again if played through a loudspeaker in the room of the remote team. The result of this double reverberation is filtering that produces a signal which sounds like it is coming from the bottom of a barrel. We were looking for a means of avoiding this rain-barrel effect. We thought the best approach was to build a microphone that would reduce reverberated sound in the source room from being received. One way to do that would be to make a directional microphone that preferentially received sound from a location at roughly the height of a seated person's mouth.

Bell Labs had worked on this problem and developed an array of 28 microphones put together in a vertical line. The outputs of each microphone were summed with appropriate weighting for each, generating a microphone system that preferentially received sound perpendicular to the line at its midpoint. While some reflections off the side walls would still be received, sound from the floor, the ceiling, and much of the side walls would be suppressed. The system design required selecting microphones that were nearly identical in behavior and adding amplifiers to each (or to each pair). Because of the sophistication of this design, it was an expensive device to construct. It was to be sold in the early 1980s at a cost of about $1000.

Ilene and I figured out that instead of 28 small microphones in a line, we could make a single long thin microphone with the same dimensions as the array. The question was simply what we could vary as a function of position in a manner equivalent to the weighting of the array amplifiers. The answer led to five patents—we could vary the shape of the electrode, the charge density, the air gap thickness, the foil thickness, or the stiffness. Of these, the simplest was to vary the electrode shape on the electret film and to place the electret directly onto a roughened surface. We built this microphone and tested it in the anechoic chamber. It worked beautifully. Compared to the array of microphones design, this design was orders of magnitude less expensive to build and introduced less electrical noise. We were awarded a Bell Labs Patent of the Year award for this work.

There were a couple of interesting things about this work on the microphone with sensitivity varying with position. First, it worked the first time we tried it. That's not the normal result in research. We anticipated needing to figure out why it hadn't worked and iterating on designs, but this one worked first time out of the chute. Second, it cemented the bond that Ilene and I formed that has persisted to this day. We approach problems from different directions, but because we try to understand each other's arguments, we make our viewpoints mesh like a Swiss watch. Over the years, my most productive periods in terms of papers have been the times I've worked with Ilene.

Ilene took the lead writing up the article on this work and was anxious to see it published. By the time it would be in print she would be gone to a university faculty position and having papers published would matter. At an Acoustical Society meeting in San Diego, Ilene and I discussed her draft of the paper, with me pushing for some additional work and Ilene resisting. We sat by the hotel pool and argued, with me saying I thought this would be a landmark paper in acoustics and worth the extra effort. Eventually I got up to find a restroom. When I returned Ilene was laughing and told me she would do the extra work. I asked what had happened and she told me that as soon as I left two of the staff from the hotel ran over and asked her if "that man was bothering me." She said that for a moment she saw how she could win the argument, but then she got indignant that they would assume the worst about me. She sent them away and caved in to my demands.

Ilene and I also worked on a project that was requested by Bill Baker, by then the former President of Bell Labs. After ABLE interacted with Bill, I had seen him as a bit of a mentor. He was supportive of our work to create opportunities for Black employees of Bell Labs, and he was aware of my work. I remember him stopping me in the hallway 1 day and asking about my family. He not only remembered who I was but knew the names of my wife and children. This was routinely one of Bill's strengths. He was able to connect with people and to remember important minutia about them.

Bill's request to Ilene and me was about Cornell Medicine. Cornell Medicine was trying to determine what single measurement of blood pressure best correlated to the 24-hour average. Bill had gotten involved in this as he had been required to wear a heart monitor to diagnose a medical problem. Because the drugs typically prescribed for high blood pressure had serious side effects, the aim was to figure out how to medicate only those needing it rather than people anxious because of being in a doctor's office (white-coat hypertension or white-coat syndrome). The equipment the study was using placed a microphone under a blood pressure cuff that automatically inflated and measured the blood pressure at regular intervals. Unfortunately, any motion of the microphone meant the signal could be lost. Bill asked me to see if we couldn't improve that.

This was my second time working with stethoscopes, the first being the stethoscope I built for Louisa's grandfather, and this time I needed to learn how to take blood pressure readings so that I could see how the equipment was working. Working with the hypertension clinic at Cornell Medical Center, we were able to improve the equipment performance. We swapped out the microphone that came with the unit for an electret microphone. This worked well but we could see that the aluminum coating of the electret in the microphone decayed from constant exposure to perspiration. A permanent solution would use a different metal coating, but for the study, what we generated was sufficient. The study found that having patients play computer games while measuring their blood pressure produced an accurate measure of the average blood pressure.

In addition to the patent recognition, I received two awards in 1982. I received the Award of Merit of the North Central Jersey Chapter of the National Technical Association (NTA) "for continuing contributions benefiting both science and community." I also received the ABLE Award "for unselfish devotion to the Association of Black Labs Employees."

11

Married Again

By the early 1980s, Melanie had completed her undergraduate degree. She started working as an assistant engineer at Sigma Sound Studios, where she had the pleasure of working with a variety of celebrities including Madonna, Roy Ayers, and Phylis Hyman. She was also doing live sound as a freelance engineer and worked on an album for Sade at Compass Point Studio in the Bahamas, formerly Bob Marly's studio. Melanie also worked with Talking Heads, receiving a platinum award for her work on the album "Little Creatures," released in 1985. Melanie had found a partner and was enjoying life in the city.

Laurie had opted not to return to school after her car accident, although she did return to school much later to earn a user interface certificate. Being more like her mother than her siblings, her interests were fixed in the arts and music, so she chose to pursue those interests.

Jay was still living at home and was making it clear he was much more interested in his music than his academics. He had become a sullen teenager cruising through life. He alternated between my home and his mother's.

Although I've never been much of a partygoer, I've always sought social engagement and connection with people. As I grew older, my values and ideas remained youthful, and I found myself naturally gravitating to people younger than me. It was a young crowd of folks that I hung out with at Bell Labs, and it was through these friends that I met a like-minded group of people in New Brunswick, New Jersey. They invited me to a party about 6 months after Kirstie and I had parted company, and I agreed to make an appearance. I had planned to leave early, but on my way out I ran into a few people coming into the party and got to chatting with them. Among this group was Marlene

J. E. West, I. J. Busch-Vishniac, *Mic Drop*, https://doi.org/10.1007/978-3-032-20922-1_11

Stevens. We spent a long time talking about her interests—mostly politics and how it resulted in the unequal treatment of Black and brown people. This is, of course, also a favorite topic of mine, having lived the disparity in a way that most people don't, because I was in a mentally competitive world and was the only Black person within my group. Marlene and I clicked, although she was clearly more politically radical than I. I think of myself as a problem solver. Marlene is more of a problem raiser. She likes to identify where the priorities for progress need to be focused.

After this party, Marlene and I started casually and infrequently staying in contact. At first, we didn't meet in person, but we talked on the phone about our passions and frustrations. At the time, Marlene was working on her master's degree in special education. As part of that work, Marlene spent a summer in Barbados at a school for special needs children. She loved the experience, including the warm weather and the slower pace of life. I worried that she wouldn't return to New Jersey, but in time she did. I was in Europe for a good bit of the summer, but I tried to stay in touch by phoning her. Marlene was living with a few women in a small building in the school compound, and they didn't have a telephone. That meant that when I called the main house, someone would go outside, ring a bell and let Marlene know that "Mr. West is calling for you." When Marlene came home, I met her at Kennedy Airport, and our relationship grew more serious.

I became far more interested in the political aspects of Marlene's life and she showed interest in the scientific part of my life. We didn't talk very much about my work except for some of my very far out dreams about things that should be done—mostly in the direction of improving the quality of life for Black people.

Marlene and I also discussed my interest in Trotsky and his movement of permanent revolution. Trotsky believed that economic systems had to be viewed as world systems rather than associated with a single nation. Thus, for real change, pressure for change of the world's economic theories was necessary.

Of course, I also thought Marlene was a very pretty lady and was drawn to her. We had an on-again off-again relationship for quite a while. We spent more time in conversation by phone than in person.

Marlene's family had lived in Newark, but after the uprising in Newark during the race riot years of the late 1960s, her family moved to a small, suburban town. Marlene's father was a lot like mine—a jack of all trades and master of a few. He was a member of the Cree nation in North Carolina and did everything from gardening to construction work. Marlene has two brothers and two sisters. One of her brothers is a police officer, possibly the only policeman I've gotten close to, as I was taught to walk in the opposite direction when I saw a policeman coming. As a child, Marlene attended Catholic school, but she changed her affiliation to the Baptist church for political reasons as an adult.

Marlene and her siblings routinely gathered for Sunday dinner at her parents' home. Eventually, she brought me along to meet her family and it didn't go particularly well. I've never paid much attention to how I dress as I put comfort above style. Once I was gone, Marlene's parents and siblings asked her why she was dating a homeless person. I guess my clothes and my slow speech cadence did not show me in the best light. I am happy to report that I eventually won over Marlene's mom. Over the years we grew close and had long conversations. Marlene's father remained a man of few words, and I never could tell how he felt about me.

At this time, I was spending summers in Europe and broached the subject of Marlene coming with me. She said she would only agree to that if she were my wife. I had avoided proposing to Marlene because I was afraid that the more than twenty-year age difference was simply too much, but with this encouragement, I asked Marlene to marry me. She agreed and we were married before the summer trip to Germany.

It's probably just as well that we married prior to the Germany trip, because Marlene did not have a good experience there. I was working nonstop, and Marlene was finding that Germans had little experience with Black people. They tended to assume she was the wife of a soldier stationed in Germany rather than a person with skills and a career of her own. She was generally treated with disdain.

Our trip to Germany did produce a great result though—Marlene was pregnant by the time we returned to the US. Coincidentally, Melanie was pregnant at the same time, so I found myself about to become a grandfather and a baby's father pretty much simultaneously.

Marlene was generally distrustful of medicine and traditional doctors. I believe this stemmed from Marlene witnessing her mother's experience with doctors. Marlene's mother had rheumatoid arthritis, a crippling disease, and the doctors were either unable or unwilling to help her overcome the pain and limitations imposed by the illness.

Marlene aimed to do a home birth and started a class with a midwife. I went to one session of the class and came away convinced the statements made and procedures used were not scientifically sound. Through my collaboration with some of the Cornell Medical School folks in New York City, I had access to their medical library. I spent a couple of days combing through the research literature to check on statistics that I believed were misquoted by the midwife. I was therefore armed at the next meeting of the midwife's group, and I aggressively corrected misstatements as they occurred. The net result was that we were kicked out of the midwife's group of patients, a result that made me happy but was not what Marlene had wanted.

Adding to the tension when Marlene was pregnant, I became concerned that my casual cocaine use was now out of control and bordering on addiction. I needed help getting off the drug, as going cold turkey was proving too difficult. I moved back in with Kirstie for a bit, and she helped me wean off the cocaine. I did return to Marlene once I was clean, but it was clear that Marlene would have preferred me getting help from anyone other than a former lover.

I was able to connect Marlene to a group of respected Black doctors I knew, and not a moment too soon. When only 7 months pregnant, Marlene went into labor while we were walking in a park. We rushed to the hospital and our beloved daughter was out of the chute within a few hours. We named her Ellington after the great jazz musician Duke Ellington. Being 2 months premature, Ellington needed to be in a neonatal intensive care unit (NICU) since her lungs were not well developed. The hospital in which Marlene had delivered Ellington didn't have a NICU, so Ellington ended up in one hospital while Marlene was in another. My job, besides trying to be the calm father, was to be the milkman—to carry the milk pumped by Marlene to the NICU for Ellington.

At first, it was tough to see Ellington because she had tubes in her and equipment surrounding her. I could hold her in the palm of one hand and use my other hand to take pictures of her to show Marlene. Ellington had a lengthier stay than Marlene, but she pulled through like a champ. Day by day Ellington grew stronger. Today, my daughter shows no signs of the fragility of her birth.

When Ellington came along, I was a much older father than for my other children. I was more established in my work and had achieved many of my life goals. I had been largely absent for much of the early life of my older children, counting on Louisa to run the household while I focused on my career. When Louisa became ill and I needed to take over, I mostly saw taking care of them as a chore. We had fun and loved our time together, but in the back of my mind, I believed I was taking time away from my work.

I was determined that this time it would be different. Now I had a partner who was an excellent mom, and I was secure financially. I would spend more time with Marlene and Ellington and be a more active influence at home. When traveling to conferences, I tried to bring Marlene and Ellington with me. If I went abroad in the summer, I also tried to bring them with me and to spend time in their company rather than working nonstop. It created a very different family dynamic. I enjoyed raising a child for the first time. Ellington and Marlene are shown with me in Fig. 11.1.

About a year after Ellington was born, Marlene learned she was pregnant again. While she was looking forward to expanding our family, I was thinking about my age, my life goals, and my work, and I insisted that we not have a second child. Ellington was my fourth and I did not want to produce a fifth. Very reluctantly, Marlene agreed to an abortion in what she regards as perhaps the decision she most regrets in her life. This decision certainly drove a major wedge between Marlene and me.

By the mid 1980s Marlene and I were ensconced in a home with baby Ellington and enjoying life as a family. We would regularly attend Sunday dinners at the home of Marlene's parents, which also let us keep up with Marlene's siblings and their families. Marlene was teaching and I was working

Fig. 11.1 Marlene and me with baby Ellington

at Bell Labs. We had more than enough money to keep us comfortable. The one difficult aspect of our lives was the interaction with my older children.

The birth of Melanie's son, Aaron, was the end of Melanie's work at Electric Lady Studios, because musicians tend to record very late at night, when the surroundings are quiet, and that was not conducive to raising a child alone, as Melanie's partner was no longer in the picture. She had, by that point, achieved quite a bit of fame for her work, and continued to freelance on interesting projects.

Melanie's sound engineering skills were so strong that I brought her into Bell Labs to work in my laboratory on a temporary basis. As other researchers became aware of her talents, she would get shanghaied by them for work in their lab on speech recognition and synthesis. She certainly was able to improve the quality of recordings of speech that were made—usually reducing the noise significantly. It was through this work at Bell Labs that Melanie met Thrasyvoulos (Thrasos) Pappas, a visual perception research scientist at Bell Labs. Thrasos and Melanie worked on an experimental teleconferencing room, with Thrasos in charge of the video and Melanie the audio. They argued constantly and, in the process, their professional relationship turned personal, leading to their marriage. Ultimately, Melanie was forced to leave Bell Labs because of nepotism rules.

In the mid-1980's, Laurie was working in art and landscaping while continuing to play music with Jay, and Jay was finishing up high school. Jay went off to Union County College with a major in electrical engineering. He dropped out after only a year to focus on the music side of his life. When a disagreement with Bernie's manager put an end to Room Service's participation in Parliament—Funkadelics, he floated for a bit and then returned to school. He did 2 years at Union County College in business and then transferred to Keene University, earning a BS in management science.

Louisa and I had always been fairly permissive parents. We would rather err on the side of allowing the kids to have too much freedom than to potentially stifle their creativity and chances to develop their own identities. But whether the result of resenting Jim for having new relationships or simply being themselves, the interactions Marlene had with my kids led her to feel disrespected. My kids spoke to Marlene, and sometimes to me or to other senior family members, in ways that Marlene would never have spoken to her parents.

Things came to a head between Laurie and Marlene at the time of Melanie's wedding to Thrasos. Melanie decided she wanted the wedding to be at my house, and she asked Laurie to handle the food. This led to a disagreement with Marlene, who wanted the food to be outside, but Laurie objected due to the heat. The disagreement escalated with the perception by Marlene that

Laurie was using a very nasty tone with her. Marlene even went so far as to call Louisa and tell her that they needed to work together to ensure that Laurie understood she couldn't talk that way to her. Louisa made excuses for Laurie rather than agreeing to intervene, and a schism between Marlene and Laurie grew that lasted for quite some time.

As Ellington got older, I proposed that Marlene home school her so we could travel together more often. I knew that this would keep me in front of Ellington frequently and allow her to experience the world outside of Plainfield. Marlene objected on the grounds that it would not be a good social and developmental solution for Ellington, and I lost that argument. This meant that while Ellington and I grew very close, it was difficult when I left on a trip. Early on, when she was only 3 or 4 years old, she would revert to baby behaviors when I travelled. Instead of going away for weeks or months at a time, I learned to commute back and forth in the summers to keep my away time limited and to take Ellington and Marlene with me as often as possible.

When Ellington was about 5 years old and struggling to learn how to read, Marlene turned the newspaper upside down and said that many of her students told her they could read better that way. Ellington looked at the paper and said it was much easier to make out the letters and read that way. I looked at the paper and experienced the same thing. Marlene pronounced both of us dyslexic. It was the first time I'd been told that dyslexia is the name for my reading problem.

As the 1980s started, all my children were living nearby, and grandchildren were arriving. It was wonderful to be surrounded by family and to easily have gatherings. Melanie was living just around the corner, and Laurie not much further away. Jay was with me or with his mother. Everyone was in Plainfield. Our house became the gathering place which suited Marlene and me just fine. Christmas morning, everyone would come over and spend the entire day munching on food from the Jewish deli nearby.

Melanie decided to pursue a Master's degree at New York University (NYU) and that led to her son, Aaron, often staying with us Monday through Friday. Of course, when Marlene and I traveled, Aaron and Ellington would both stay with Melanie. Having both Aaron and Ellington in the house meant they became great companions. This arrangement continued from kindergarten through fourth grade. It also led to great experiences for them. I know that Melanie took Ellington and Aaron with her when she worked on a project at Disney's EPCOT Center. They roamed the park all day while Melanie worked.

On Saturday mornings, I would wake up the kids and make either garlic bread from the bread purchased from our local bakery or pancakes for breakfast. I was doing a fair amount of cooking at this time, more often than not, Italian meals of some sort. After breakfast and cartoons, we'd go grocery shopping. The kids loved this as I would let them put anything within reason into the shopping cart. As we drove around town, we'd sing along to the music playing in the car.

When Ellington and Aaron were in a soccer league for kids 7 years old and under, Melanie's husband, Thrasos, and I were the team coaches. We thought this was a great opportunity to teach the kids to learn to compete and to never give up, but several of the parents thought we were being too competitive and needed to tone it down. Our tenure as coaches was cut short.

For part of the time Aaron was living with us, my half-brother Leroy was also living with us. Leroy had struggled after returning from his time in the Army and seemed to have inherited our father's taste for alcohol. Both he and Sonny were alcoholics. Sonny was living a screwed up adult life in Washington DC until drinking took him down. Leroy had been living in New Jersey and had put together a successful construction business working as a mason, but I got a call one day from someone who knew him saying that he was sleeping in a derelict car in Newark because he'd lost his apartment. I found Leroy and got him to the VA hospital. I let him know that he had a place to stay as long as he was sober, and he took me up on the offer. Upon leaving the VA hospital, Leroy came to live with us. He is shown in Fig. 11.2 along with his son, Ricky, my son, Jay, my brother and me.

The Army had made Leroy regimented and rule-driven, and he became another father figure to Ellington and Aaron. This seemed to give some purpose to his life, and he was able to stay sober … mostly. On those occasions when he would fall off the wagon, he had enough control to climb back on of his own accord. Leroy stayed with us until Ellington was in eighth grade, at which point he moved in with his girlfriend. Alas, Leroy died just 2 years later.

There are lots of stories I could relate about Ellington and ASA meetings but let me mention just one here. At a meeting in Hawaii in 1988 that was joint with the Acoustical Society of Japan, we lost Ellington, then about 6 years old, for a short while. She and Marlene had come down in the elevator to the lobby, but Ellington walked back into the elevator when Marlene had her back to her, and she was gone. Fortunately, many ASA members knew who Ellington was at that point and we started looking on every floor for her. She was fine and having a great time exploring the hotel. We found her after just a few minutes and kept a closer eye on her for the rest of the trip.

When Ellington was still quite young, Take Your Daughter to Work Day became a thing in the US. I would bring Ellington with me for the day, and she would have a blast. Bell Labs always arranged for a series of activities and

Fig. 11.2 West men: from left Ricky (Leroy's son), Jay, me, Nate, and Leroy

demonstrations on these days and Ellington loved it. Apparently, she would go to school the next day and tell her friends about all the awesome things she saw, making them very jealous.

For my older kids I pretty much never attended any of their graduations or dance recitals or art shows or similar. I would ask what these events were doing for them long term, especially graduations, which were merely acknowledging that they performed what was expected of them. This is apparently an attitude I got from my family, where rewards were hard to earn. But in Marlene's family, little successes were deserving of note and reward, so for Ellington, I attended every graduation and even sat through dance recitals.

I've always led a very physically active life, and at this time I took up racquetball. There was a group of Black men from Bell Labs who took to the game as it grew in popularity, and we would gather regularly to play at various courts. I recall that trying to avoid a particularly good shot coming right at me, I jumped and then lifted one of my legs to avoid the ball hitting me. I came down hard on my left leg and blew out my meniscus.

In the mid 1980s I started having some trouble with my back. I am skinny and don't carry a lot of weight, but I had managed through various activities in my life to create some spinal disc problems. This is the sort of problem that doesn't get better on its own with time and it certainly didn't for me. I've never been a supporter of unnecessary surgeries so I decided to ignore the problem for as long as I could. Ultimately my doctors recommended putting a rod into my spine. I asked around and read what I could find and determined that most of the people who had this surgery weren't happy with the outcome, so I declined to have the surgery. This was the beginning of my loss of mobility, a serious problem that has defined a good bit of my final years.

The 1980s marked the start of my second family, with my marriage to Marlene and the birth of Ellington. It was a time of great excitement on the home front.

12

Digital Signal Processing

In the early 1980s, while attending an ASA meeting, I was approached by a senior acoustician at Penn State University, Jiri Tichy, asking me to consider hiring a student he had who was finishing his PhD. Jiri made it clear that he didn't understand what made this student tick but that he was very bright and had the potential to make huge contributions in acoustics. Jiri asked me to meet with him which I agreed to do. The way Jiri was talking about Gary Elko I was expecting some nerdy guy who was socially awkward and who couldn't chew gum and talk at the same time, but in walked a young man with long blond hair, dressed like a hippie. I talked with him for a couple of hours and then invited him to Bell Labs to give a talk because it was clear he was very smart and the work he was doing was cutting edge. Gary did come to Bell Labs for a presentation, and he was ultimately hired by my department, joining in 1984.

Gary opened up a whole new world for me. Up until that point, everything I had done had been based on traditional electronics. Gary came in with a knowledge of digital signal processing, and he showed me what it could do. In traditional electronics, one uses voltages or currents as signals and their characteristics, such as amplitude or frequency content, are what determines whether the process is producing desired results. In digital signal processing, real-world signals are converted to digital data (a series of ones and zeros). This permits unique approaches to compress, filter, and manipulate the signals. With Gary's knowledge of digital signal processing techniques and my expertise in sensors and actuators, we developed a strong partnership.

At the time Gary joined Bell Labs, the organization was in its heyday. Researchers from around the world would come and present their work in

J. E. West, I. J. Busch-Vishniac, *Mic Drop*, https://doi.org/10.1007/978-3-032-20922-1_12

talks held in the Murray Hill, NJ auditorium. But Bell Labs was massive, with people in Chicago, Georgia, Holmdel, NJ and elsewhere, so the presentations at Murray Hill were broadcast to remote locations. I know that we do that now almost without thinking about the technology, but in the 1980s, connecting remote locations was not easy. The acoustic quality of the broadcast was problematic in these presentations, particularly if questions were asked in Murray Hill. Remote listeners would have to ask the presenter to repeat the question as the questioner was not miked, and inevitably, presenters would forget to repeat the questions. In part, the problem was caused by the culture at Bell Labs, Murray Hill. It's not too hard to remember to repeat the questions when they all are held until the end of the presentation. Murray Hill scientists were always rude and would just shout out questions during a talk and expect the speaker to respond immediately. Frequently, the speaker would mention that they were just about to get into that issue in their talk if the audience would let them continue.

Gary and I talked about this problem and came up with a solution: a rectangular array of sensors that could hang on the front wall of the auditorium and pick up sound from the audience. The processing would find the speaker in the audience and then selectively record sound from him or her for broadcast. Gary took on building this 10 × 10 sensor array, which was an enormous amount of work. I was very impressed with his determination to get this done. When he was finished and had tested it, it was installed in the auditorium. For all I know, it might still be there. For me, this was a new approach to directive sensors. I had previously worked on individual microphones that were directional by design. This project was using the other approach to making a directional sensor—using many microphones combined to produce a response in favored directions.

Gary's array of microphones was, in actuality, three arrays combined in one physical structure. The central array with close sensor spacing was used for high frequency sound. The array including some of the central microphones and those beyond the high frequency array was used for mid-frequency sound. The spacing between microphones in this array was greater than for the high frequency array. The outer sensors and specific sensors interior to the array were used to create an array with microphones spaced even further apart. This was used to pick up low frequency sound. In this manner, Gary's harmonically nested array was able to pick up sound well over the entire speech band, from roughly 50 to 5000 Hz. Through measurement, we determined that it was at the very high frequency end of this array set that the system had challenges, but fortunately, this was outside the speech range and thus not terribly interesting.

Building this array was an immense project. For the array to work, we needed to use microphones whose response characteristics were nearly identical. At one point Gary and I took a trip to the Primo factory in Japan, where they made good quality electret microphones. We wanted to see how the microphones were built and to understand the likelihood that we could sort them and find a group whose response was almost constant over a broad frequency range. Much to my embarrassment, during that trip to Japan I learned that Japanese women found my looks very exotic, and they kept staring, giggling, and even approaching me and flirting. Apparently, I was one of the first Black people to find my way to that Japanese location.

I am eternally grateful to Gary for introducing me to digital signal processing. It changed the course of my career and allowed me to continue to be productive as an acoustics researcher. Although we haven't worked together for some years, we continue to stay in touch occasionally, mostly because we both enjoyed interacting on a professional and personal level.

The mid to late 1980s continued to be times when I received recognition for my activities. In 1985 I was made a Fellow of both the ASA and the IEEE. While I appreciated both these membership upgrades, the IEEE Fellowship led to an embarrassing event. The New Jersey Chapter of the IEEE threw a reception in my honor when I was named an IEEE Fellow. Marlene and I showed up for the dinner and were seated at the head table with the New Jersey Chapter President and his wife. Unfortunately, the wife of the chapter President, who engaged us in conversation as soon as we arrived, seemed unable to accept that Marlene and I were Black. She asked us repeatedly about our background and assumed we were from India. When we corrected her, she was simply unable to accept what we said. Exasperated and offended, we left the dinner before appetizers were served.

In 1985, I received the Outstanding Performance Award of the Grant Avenue Community Center Board of Trustees. The Grant Avenue Community Center was the premier Black social services agency in Plainfield. It was later one of the Black United Fund organizations. My community efforts were also recognized in a tribute by the Honorable Donald M. Payne, a US Representative, in 1988. It was gratifying to have my community activities noticed and rewarded.

My efforts with Bell Labs to improve opportunities for Black employees were also recognized in the 1980s. In 1988 I received the AT&T Bell Labs Research Affirmative Action Award, and in 1989, the Lewis Latimore Award of the organization now known as NETwork BICP.

Digital signal processing is now the center of my research work. It was introduced to me in the 1980s by Gary Elko and quickly became an important tool in acoustics, as evidenced by the growing number of talks using digital techniques at the Acoustical Society of America meetings.

13

Still Parenting

Ellington likes to tell a story about hiking that dates from the early 1990s and that she thinks describes my parenting style. I am a walker and always have been. On weekends, I would take whoever was around on long walks through the forest or a park or neighborhood. I spent lots of time trying to make sure the kids could navigate on their own, partly to ensure they were safe and partly to instill confidence in them. I taught them how to read a map and to look for signs on a trail. One weekend, Marlene and I went walking in the woods with Ellington and Aaron, who were in front of us. We quietly left them and returned to our car, leaving them to navigate alone. They told us they first panicked and then worried that we were hurt before they realized we were just testing their ability to find their way back. It took some time, but the two of them made their way to us and let us know they were really pissed at us for leaving them.

Truly, this is indicative of my parenting style—Ellington is not wrong. I believe in teaching and providing resources and then setting up situations that use that learning and instill confidence, even if it feels like being thrown into the deep end. Mostly, Marlene agrees with me. The only difference we have on this is on degree of challenge it is appropriate to present. And Ellington and Aaron knew that we weren't far away. We would have been able to rescue them if necessary.

By the 1990s, my older children were doing well. Melanie had completed a Master's degree in interactive telecommunications from the Tisch School of the Arts at NYU. Jay was working for Breeze-Eastern, a company that designs and manufactures helicopter rescue hoists, winches, cargo hooks, and weapons handling systems. He started as an electrical assembler but quickly moved

J. E. West, I. J. Busch-Vishniac, *Mic Drop*, https://doi.org/10.1007/978-3-032-20922-1_13

up to become a project engineer and finally a quality assurance engineer. In 2025 Jay celebrated being with Breeze-Eastern for 35 years.

Laurie had met Mickey Davis while they were both working for a landscape company, and they married in 1993. Mickey later shifted to work as the manager of the Short Hills Mall in Short Hills, NJ and Laurie to work in the operations department of a company that provides a marketing service for real estate firms. When Laurie left to live in western New Jersey, she was the first of my kids to leave the seat of their upbringing. Melanie and Thrasos left shortly later, landing in Chicago. Their daughter, Kaliroë, was born a year after Melanie and Thrasos married.

Together, Laurie and Melanie cofounded Tiz Media Foundation in 2003. They worked with youth across the country and in Canada to create multimedia STEM learning modules, called MindRap, using music art, and storytelling to explore topics such as climate change, neuroscience, and the physics of sound. They were supported in part by the National Science Foundation and Motorola.

In 1993, Jay married Jackie Brathwaite, a woman he had met in tenth grade, and reconnected with many years later. They had son James in 1996 and son Julian in 1998. Jay still has studio equipment in his home and continues to work on his music. Both of the boys are interested in music as well.

Louisa passed away in the 1990s and our kids inherited her house. By this time Louisa had long since stopped abusing pharmaceuticals and alcohol. The kids had known about Louisa's drinking and bouts of depression, but I am not sure they knew about her abuse of pharmaceuticals. The kids had a relationship with Louisa until the very end, and I didn't want to prevent that from happening. I also wanted to protect Louisa's family to some extent. I am the Godfather to Ian, the son of Louisa's brother, and I like the whole family. The kids decided to sell Louisa's home, but they didn't choose Marlene to be the real estate agent even though she was managing a real estate company with a partner at that time. Instead, they asked her business partner to list the home. It was impossible for Marlene to see this as anything other than a personal insult. The kids could have used dozens of other real estate firms, but they pressed to use her firm but NOT her. Marlene and her partner ultimately agreed that the partner could handle the home sale, but this meant the kids would have to pay the seller's commission, something Marlene would have waived for them.

Having Marlene closer to the age of my older children rather than to me has made it hard for her to command the respect that is due her. While this continues to be a problem and causes occasional awkward moments at family gatherings, everyone has made their peace with the situation at this point.

Similarly, Ellington often sees her half-siblings more as parental figures than as sisters and brothers given the age difference.

In addition to the strain put on our marriage by my children's treatment of Marlene, my infidelity became an issue that divided us. I have always had a rather liberal view of sexual relations, and I've been called a player by many people, including my children. I love Marlene and was working hard to support the family, but as a rather handsome and charming and successful Black man, I had many women offer their services to me, and I honestly didn't always turn them down. I confess that I am simply not temperamentally suited to committed monogamous relationships. I tried to keep my indiscretions hidden, but Marlene noticed that I had ceased coming to Sunday dinners at her parents' house. I was using that time to see other women, and the discovery of this devastated Marlene. Marlene was ashamed but didn't want to tell her mother and ruin the relationship I had with her. She didn't want Ellington to be a child of divorced parents, so she stayed with me, but our relationship was changed. Gone was the trust that had previously existed.

The 1990s were also the decade in which my mother passed away. This was a very painful part of my life. Mother had lived alone for years in Virginia, but I eventually convinced her to move to New Jersey to be close to my family. She moved into an apartment, and we included her in family events. I recall that Mother grew to know and be fond of Marlene's mother, the two of them becoming our beloved mothers-in-law. But eventually, Mother's memory began to fail her. She had episodes of turning on burners and forgetting that she had done so and starting to run water for a bath and forgetting that the water was running. Her apartment complex decided she couldn't safely stay there any longer and we needed to find a place for her. After much searching, we found a place for Mother in a memory care unit in Plainfield. I hated having to remove her from family surroundings as it felt like an abandonment, but we simply couldn't care for her on our own. Even in the memory care facility, with its many staff and its locked doors, she was occasionally escaping. In the end, Mother died when she was 97 years old. Her passing marked my transition into the oldest generation of the family.

Shortly after my mother passed away, I took Ellington with me for a trip to Farmville. I wanted her to have a chance to see where I had grown up and to meet some of the relatives I still had there. I introduced her to Edwilda and Baby Sister and showed her the dirt roads and tiny commercial area that defined the town. Ellington was left with the impression that life moved much more slowly in Farmville than in Plainfield and that community was the center of activity there. She also learned that my family had long been progressive—I hadn't strayed from the norm in developing my ways. And as a fun treat, I let Ellington have a chance to drive on the nearly deserted dirt roads near my former home even though she was only 10 or 11 years old. As I had hoped, this trip is a memory that has stayed with her.

14

I Am the President of the Acoustical Society of America

The decade of the 1990s was one in which a good bit of my personal and professional life centered around the ASA. I attended their biennial meetings, gave talks at the conferences, served on committees and published in their journal.

In 1991, the ASA met in Baltimore and Marlene and Ellington came with me. During the day, Marlene was taking in the sites in Baltimore with Ellington while I attended the conference. One day Marlene and Ellington took a cab to the American Visionary Arts Museum in Federal Hill. When it was time to return to the hotel, Marlene found that not one cab would stop for them. Presumably, the cab drivers were making assumptions about a Black woman with a young child and willingness to pay or tip. Either that, or they just didn't want to dirty their cabs with Black people. Eventually my family made their way back to the hotel, walking for a good bit of the way through some sketchy areas. When they arrived back at the hotel and told conference organizers what had happened, the ASA filed a complaint with the city. The fact that the ASA staff were so indignant about Marlene's treatment endeared them to both of us. Unfortunately, it did not present Baltimore in a good light, and this came back to bite us later.

As my career blossomed, my involvement with the ASA grew. I attended their meetings twice a year and organized a few special topics sessions. I always felt welcomed by the members but couldn't help noticing that there were a vanishingly small number of Black people attending the conferences. I suggested that we needed to do something to make at least a modest attempt to diversify the society to make it clear that students of color were welcome, and that acoustics was an accessible avenue of pursuit. I suggested that the ASA

J. E. West, I. J. Busch-Vishniac, *Mic Drop*, https://doi.org/10.1007/978-3-032-20922-1_14

fund a fellowship for minority students, and the society management agreed to set this up. The Minority Fellowship was established in 1992 and the first fellowship awarded in 1993. The fellowship provides support for 2 years to a recent PhD graduate and has brought wonderful new acousticians into the fold. I have tried to get to know these young acousticians. I have been particularly concerned that these wonderful young people be seen not as poor, underprivileged kids, but as equals to all the society members. In 2018, the Minority Fellowship was renamed the James West Fellowship in my honor. As of 2025, 18 scholars have held the Minority or James West Fellowship.

In 1996, I was asked to be a candidate for President of the society. The position is effectively a three-year term in office: a year each as President-elect, President, and past President. If elected, I would be the first officer of color in the history of the organization. Because of my love of the ASA and my desire to push for greater diversity, I agreed to stand for election and ended up pitted against Patricia Kuhl, who would be the first woman to serve as President if elected. I won that election and began service as the President-elect at the Spring meeting of the ASA in 1997.

The ASA works in the same ways many scientific professional societies operate. It has a small set of paid staff members who handle the bulk of the administrative work and a volunteer board to set policies. As with most boards, a great deal of the work is mundane and boring. I set out to have some impact beyond continuing the norm. Specifically, I accomplished two important goals while serving as President of the ASA. I replaced the Treasurer, and I pushed for our first official stand-alone conference outside of the US.

When I first joined the ASA Executive Council as the President-elect, I realized that a good bit of the agenda dealt with funding. I was having trouble understanding how we were doing financially and asked the Treasurer, Bill Lang, for a report. Although the job of Treasurer mandated regular reports to the Council, Bill had not filed a formal report for a few years. My polite request produced no response, so I requested a report again and set a deadline for having it delivered. At that point, Bill told me that my request sounded like an order. I responded by saying he could certainly hear it that way. At that point, Bill told me he didn't take orders from niggers, making it very clear what I was dealing with. The Treasurer position at that time was not filled through election but rather through appointment by the Executive Council. It is one of the few paid staff positions. I went to the Executive Council and said we needed to replace the Treasurer as he was shirking his responsibilities. My request was approved with no resistance and the process for hiring a new Treasurer started. It was ultimately filled by Paul Ostergard, a consulting business owner who was much more forthcoming with information.

The ASA normally has two conferences a year—one in the Fall and one in the Spring. These are held in a North American city, typically where there is a significant cluster of ASA members who volunteer their time to organize the meeting. Every third year, there is a meeting of all the acoustics professional societies in the world and these are often held in coordination with the ASA, even when outside of the US and Canada. But the ASA had never organized one of its stand-alone meetings outside of North America even though a significant percentage of its members were from China, Europe, and Mexico. After significant pushing on my part, I got the ASA to agree to a meeting in Cancun, Mexico. That meeting took place in 2002 and had a high attendance. It set the tone for the future, as the ASA now routinely makes overtures to members outside North America. Indeed, the President in 2023–2024 was a German acoustician, the first President of the ASA from outside North America.

The planning for the Cancun meeting took a great deal of work and coordination. At one point, when I was headed to Mexico to participate in a planning meeting, I decided to take Ellington and her friend, Laura, with me. The girls were about 13 years old at the time. The planning meeting was being held in Puebla, the capital and largest city in the state of Puebla, a little more than 150 miles from Mexico City. When we got to the airport in Newark, we were stopped. I was asked, "Do you have a notarized document that lets you take Laura out of the country? She is not your child, sir." I did not as we hadn't anticipated this problem.

So, the girls got into a cab and went to Marlene's office to get a notarized letter. Fortunately, Marlene is a certified notary public. Meanwhile, I went on to Mexico City and then Puebla without the girls. I gave them my credit card and told them to book a different flight to Mexico City, to find a hotel to spend the night, and then to text me as I would come pick them up. Ellington thoughtfully, and with my direction, didn't tell Marlene that I was flying without them.

The girls were able to get onto a later flight to Mexico City and when they landed, they managed to get a cab and asked to be taken to an Intercontinental Hotel since that was the only brand name they could come up with at that moment. They asked for a room and the staff were startled. They didn't know how to handle minor children traveling alone. It took some doing to convince the hotel to give them a room. The staff asked them to leave their passports, but the girls declined, asking them instead to take copies of their documents. They didn't know at the time that leaving your passport with the front desk was standard in most countries.

The girls had a lovely time in the hotel if the room service bill is any indication. In the morning, they decided to walk to a local shopping center, but they got lost on their way home. Ultimately, they arrived back at the hotel 2.5 hours after they were supposed to meet me. I was frantic by then. Had they been abducted or hurt? Why weren't they here? I really ripped into them for being so irresponsible. Ellington got even with me though for giving her a hard time. Once we got home, she told Marlene that I had gone on without them and left them to fend for themselves. Even though it was clear Ellington and Laura were fine, Marlene was livid with my abandonment of the girls. I'm not sure Laura's parents were ever told of my so-called abandonment.

The Cancun meeting was not the only meeting of the 1990s that was outside North America. In 1999 the ASA meeting, while I was President, was held in Berlin in coordination with the International Congress of Acoustics. A few of the officers brought families with them and we ended up putting the four children of the officers in a hotel room together: Ellington, Ilene's two daughters, and the daughter of Patricia Kuhl. The kids had a wonderful time even if they did live more up to the style of a rock band than polite children. The hotel did not complain but refused to clean the room while they were occupying it. Evidence suggests the kids lived on junk food and soda that week. I know that Ilene ran into credit card issues when she went to check out and ended up telling the hotel she would have to call them the following day from Baltimore with the number, as Mastercard was down in Europe. The hotel was so anxious to get rid of the kids that they just let her go with her family. They took her business card, and she did make good on the bill the next day.

I am proud of what I accomplished with the ASA. Foreign meetings continue to be sprinkled into the agenda, and a focus on diversity remains. The ASA hosts a blog for Spanish-speaking acousticians and has a committee working on how to increase the number of Black acousticians. The postdoctoral fellowship named in my honor continues to fund talented Black students in the field. When I attend meetings, mine is not the only Black face at the conference, although there are still few enough to count on just my hands.

15

Concert Halls and Microphones

The 1990s were a time when I continued to have a lot of fun at Bell Labs looking into research questions of interest to the company and to me. This included work on heart sound monitoring, on hands-free communication, and on a close talking microphone. I also continued my work with Gary Elko.

Gary's work on the microphone array was just one of the projects he and I worked on together. Additionally, Gary and I took on a couple of projects that related to room acoustics. Thanks to my work earlier on Philharmonic Hall in New York City, I knew that concert halls seem to show sound distortion at low frequencies, although the cause had never been articulated. The source was clearly something in the room, but it seemed to be subtle. Computer models of the room acoustics agreed fairly well with the measurements except at these low frequencies. Gary and I speculated that the problem related to something happening with reflections of sound in the room. We tested the diffraction (wave redirecting and expanding) off various shapes in the anechoic chamber thinking this might explain the issue. Eventually, I concluded that the distortion was due to sound raking off the backs of the seats in the hall.

Concert hall seats are intentionally made to be absorptive acoustically. The aim is for them to mimic the absorption of a human so that the sound is pretty much the same whether seats are occupied or empty. Seat tops and backs are necessarily hard to support the structure of the chair, and thus more reflective of sound.

A musician looking out from the stage sees a row of seats in front, and another behind it by about three feet, and another three feet beyond that, etc.—in other words a line of seats that repeats every three feet. The general rule in acoustics is that such a periodicity defines a fundamental frequency

J. E. West, I. J. Busch-Vishniac, *Mic Drop*, https://doi.org/10.1007/978-3-032-20922-1_15

which corresponds to the distance between features being a half wavelength. Given the speed of sound in air is 1125 feet per second, this means the seat rows represent a fundamental frequency of 188 Hz. If the seat tops are reflective, then the reflections will reinforce the sound at 188 Hz and its harmonics (integer multiples of 188 Hz). If the seats are sound absorbing, then the sound will show drops in the seat spacing frequency and its harmonics.

The remaining question is why this disproportionately affects low frequencies since the harmonics span from low to high frequencies. There are two reasons for this. First, the materials we use in concert halls are much better absorbers of sound at high frequencies than at low frequencies, so the higher harmonics tend to disappear more quickly. Additionally, it's important to remember that our perception is logarithmic rather than linear. The harmonics increase additively but the octave bands expand multiplicatively, so as we rise in octave bands, we have more and more harmonics in them, which tends to smooth out the perceptions of the peaks and valleys in sound.

Gary and I hit it off in more than just professional ways. The young man who Jiri Tichy couldn't understand had lots in common with me, so perhaps that's what prompted Jiri to suggest we meet. Neither of us would ever be caught in a suit if it could be avoided. Both of us enjoy music and good food. Politically we are similar in view. We became good friends and socialized outside the lab. This was not my norm and speaks volumes for how much I enjoy Gary and his family.

On a lark, Gary introduced me to Bill Nye the Science Guy, who had been one of his roommates in college. I believe it was that interaction that led to my being chosen later as one of the scientists to be featured on a Way Cool Science trading card. Bill Nye has been a strong backer of science and took on a variety of projects beyond the TV show to encourage children to get interested in the sciences. The creation of a game with trading cards was one such project. I had a copy of the card at one point, but I no longer can find it. The Way Cool Science trading cards were released by Disney in 1995 and manufactured by Skybox.

Having established a collaboration with the hypertension team at Cornell Medical School through our work on an automated blood pressure monitor, we expanded our work into related studies of blood pressure. We found that the normal noninvasive method of measuring blood pressure using blood flow (Korotkoff) sounds, tended to be very inaccurate compared to invasive methods at high and low blood pressures. This raised a good bit of concern because the patients that were most in need of treatment were those for whom the standard cuff method was not an accurate measure—i.e., people with high or low blood pressure. We considered a number of alternative measures to replace or embellish the standard measuring method, but none stood out as very helpful.

In a second study with the Cornell Medical School team, we looked at the accuracy of cuff blood pressure measurements compared to invasive methods, specifically looking at the diastolic pressure (lower blood pressure number) relative to the systolic pressure (higher blood pressure number). The team found that the difference between invasive and cuff determination of the diastolic pressure was greatest when the difference between the systolic and diastolic pressures was the lowest. The treatment methods for blood pressure at the time required medicating anyone with a diastolic pressure over 90 mmHg, so errors in this measure were significant. They could lead to people needing treatment being missed and those not needing treatment being dosed with medications with side effects. I'm not sure what direction the work went after our papers.

In the early 1990s, Jens Blauert took a sabbatical with me at Bell Labs. Jens is a German acoustician (emeritus professor) whose research is primarily in the psychoacoustics area of spatial hearing. This is the study of how humans process sound to understand the space from whence it came. We worked on an interesting project on sound localization. If one records sound in the ears of a listener using tiny microphones, then plays it for a second listener using headphones, the second listener often perceives the sound image to be behind that of the original and possibly elevated from it. It turns out that this is the result of each of us having unique characteristics introduced by the shape of our ears. Jens and I decided to improve upon this situation by creating a sound playback system that was about 10 cm in front of the listener. By putting small speakers on glasses, we could then provide the sound in front of the listener and allow his or her ears to respond normally. Our work successfully demonstrated a means of preserving source localization in teleconferencing and other applications.

I also got involved in a project at Bell Labs on hands-free communication for use in automobiles. The main idea here was to enable a directive microphone aimed at the height of the mouth of the driver. We produced a few different directional microphones, but the one that most intrigued me and that I largely developed used what I called an image microphone.

One of the standard means of obtaining directivity is to use multiple microphones placed in a line. Gary had relied upon this technique in creating the sensor array for telecommunication of talks at Murray Hill. I had tended to shy away from this approach to building a directive sensor because it is very hard to find two or more identical microphones and that is what is needed for an array to work well. But I had an "aha" moment in the hands-free microphone work and realized that cars are small spaces with lots of highly reflective surfaces. If a microphone were put near to but not on the windshield, for instance, then sound would reflect off the windshield and the microphone would hear the reflection as well as the direct sound. Because the windshield is a very hard surface, effectively no sound would be absorbed upon reflection, so the strength of the reflected sound and direct sound would be the same. Then the only difference in the direct and reflected sound signals is how far they travel from the source to the microphone. We tend to model this in acoustics by replacing the reflecting surface (the windshield in this case) with an image of the microphone aligned so that the reflected path is represented as the straight path to the image microphone. In other words, we forget about the windshield but imagine there are two microphones—the original and an image beyond the windshield by exactly the distance the microphone is offset from the windshield. This image microphone is precisely the same as the physical microphone, so the microphone offset creates a pair of identical microphones in a line. No care has to be taken to sort through buckets of microphones to find a set that is nearly identical in characteristics.

When we tested a directional microphone based on this principle, it worked well. I'm not sure that Bell or any other company has capitalized on this approach for hands-free communication, but I was pleased with how this project went.

Another acoustics project I worked on during the 90 s, was a close talking microphone for use in very noisy environments. When AT&T decided it would expand internationally, it chose Formula One as one of the advertising platforms. Formula One agreed to trade technology for advertising, and communication with drivers was an issue important to them. This became abundantly clear when a racing team out of England had a serious mishap. The team monitoring the race told the driver that the passenger rear tire was sending a signal saying it was losing air, so he should take it easy. The driver, in the extremely noisy environment of the car interior, heard "fire" rather than "tire" and bailed out of the car to avoid a serious injury. He had asked for clarification, but the monitoring team could not hear him clearly. Communication was effectively one-way only, and even that was flawed.

When I looked at what they were using for communication with the driver, I saw that the helmets had a big biscuit-shaped ceramic microphone that was supposed to sit just in front of the driver's lips. The drivers were refusing to use this because they were worried that in any accident, this biscuit would smack them in the face. The first thing they did when putting on their helmets was to push that biscuit back up into the helmet body. That meant there was no microphone available for communication with the team.

What I did was use electret microphones to create first order or second order gradient microphones that were small, so they didn't mind putting the sensor near their mouth. This was better than the throat microphone option they were trying out. Throat microphones are mics that are placed directly on the skin outside of the throat to pick up the sound of speech. The throat microphone being tested only allowed drivers to use a vocabulary of 10 words that could be clearly understood. The new microphone allowed drivers to speak and be understood. I was thrilled to be involved with the racing teams and to have made a difference in a sport I loved.

The 1990s were the decade during which I seemed to endlessly receive medals and awards. In 1994, I was made a Bell Labs Fellow, a title given in recognition of years of important and impactful research at the lab. It was considered the highest technical honor given at Bell Labs. I was very proud of this award.

In 1995, I was awarded the Silver Medal in Engineering Acoustics by the ASA for my work in polymer electret transducers. In the same year the New Jersey Inventors Hall of Fame named me their Inventor of the Year. This award is presented at an annual banquet.

In 1997, I received an honorary degree from New Jersey Institute of Technology. I also was recognized by the National Society of Black Engineers (NSBE) for outstanding contributions to science and engineering. NSBE aims "to increase the number of culturally responsible black engineers who excel academically, succeed professionally and positively impact the community." It is the largest student-run organization in the US with more than 20,000 members. The following year, NSBE awarded me the Golden Torch Award. This award recognizes people who exemplify NSBE's ideals. The torch symbolizes the everlasting burning desire to be successful engineers. I was touched to receive these two NSBE awards as they align so well with my goals.

In 1998, I was elected to the National Academy of Engineering (NAE). The NAE is a private nonprofit organization that provides engineering service to the nation. New members are elected by current members, who number roughly 2000. In 2023, the NAE reported that 5% of their members identified as underrepresented minorities, an improvement from when I was inducted as one of the first members of color.

In 1999, Gerhard and I were inducted into the National Inventors Hall of Fame (NIHF). The mission of the NIHF is "to recognize inventors, promote creativity, and advance the spirit of innovation and entrepreneurship." All inductees must have at least one US patent and have invented something that was ingenious and impactful. I am shown at the NIHF reception with Marlene in Fig. 15.1.

I hadn't heard of the NIHF so when they called me, I was sure they were looking for donations. I told them "I don't have any money," and started to hang up. Fortunately, they took pains to explain they weren't asking for funds but instead telling me that they wanted to induct me. I remember this phone call being the first time I thought of myself as an inventor. I generally thought of myself as a scientist, but I hadn't made that leap to thinking about my work on applications of technology defining me as an inventor. It changed my life. It allowed me to think about the applications of my research even though I might not have made the commercial device. Identifying as an inventor also let me change how I talked to youths as I tried to interest Black children in science, math, and engineering. It was far easier to spark that interest with tales of invention and how ingenuity had changed our lives for the better, than to focus solely on the joys of finding new information in a lab.

Some years later, I had a conversation with David Fink, the NIHF Director at that time. He had interviewed a fair number of the NIHF inductees to try to gather some insight into what had made them different from others. The common theme seemed to be curiosity. I was not surprised by this result.

Fig. 15.1 Marlene and me at the National Inventors Hall of Fame reception

Curiosity has always driven my life, and I encourage others to be curious. And it doesn't have to be limited to the technical fields. It is appropriate for an artist to ask why certain shapes or colors or textures are more pleasing than others.

Also in 1999, I received the Ronald H. Brown American Innovator Award from the US Department of Commerce. This award was named for a Secretary of Commerce who served in the Clinton administration and had chaired the Democratic National Committee. He was the first Black man to hold these positions.

The decade of the 1990s was one of much activity. I assumed a leadership role in my favorite professional society and continued my collaboration with Gary Elko. I also enjoyed watching Ellington grow and to participate in the successes of my older children.

16

Leaving Bell Labs

Starting in the 1960s, early in my career, AT&T came to dominate telecommunications in the US. By the 1970s, AT&T had become the largest company in the world. AT&T was so dominant it was essentially a monopoly by the 1960s. I listened to management state that AT&T provided a service for which the operating costs were so high that it only made sense for there to be a single provider. Further, AT&T leaders argued that their use of cross-subsidies meant that basic services were accessible for all. These cross-subsidies, for instance, used higher fees in long distance calling to support lower fees for local phone calls. They also used higher than cost business accounts to subsidize home accounts. Working at Bell Labs was the first time, outside of the army, that I was "part of the system" rather than forced to be an outsider. I liked that sense of belonging to something bigger than me.

With the rise of computer technology in the 1960s, the US government sought to ensure fair practices in the increasingly important area of communication. In 1971, the FCC issued Computer Inquiry #1 and banned AT&T from entering the data processing and online services arena. I had no way of knowing at the time that this decision would have a direct effect on me and on my work, so I paid little attention to the decision. Among some of my research colleagues, however, there was very real concern that their areas of research would no longer be supported.

In 1974, the Justice Department accused AT&T, Bell Labs, and Western Electric of colluding to monopolize the telephone industry. From my perspective, I would say that we did everything possible to grow the business and knock out competition, including working together (colluding). The case didn't go to trial until 1980, and in 1984, Judge Harold Greene completed a

J. E. West, I. J. Busch-Vishniac, *Mic Drop*, https://doi.org/10.1007/978-3-032-20922-1_16

Consent Decree. The agreement released AT&T from an earlier decree that barred it from entering new markets but required it to divest its 22 local operating companies. This meant that from that moment forward, the AT&T focus on telecommunications would be long distance rather than local connections. The 22 local operating companies were consolidated into seven entities and became known as the "Baby Bells." Verizon is an example of a modern company that began life as a Baby Bell—in this case Bell Atlantic Corporation.

The 1984 Consent Decree has come to be known as the divestiture of AT&T. The local operating companies were spun off, but AT&T retained Western Electric, Yellow Pages, Bell Labs, and AT&T Long Distance. The day the decision came down, those of us working at Bell Labs could talk of nothing else. We were unsure how Bell Labs would be affected and hoped that this change would allow us to continue to think about the future of telecommunications. We worried about how lab funding and management would change given the elimination of revenue from local telephone service. We would soon find out.

With the 1984 divestiture of AT&T, Bell Labs became a subsidiary to AT&T Technologies and the funding for the Labs was significantly decreased. This immediately affected Bell Labs research programs but seemed to center outside of the Acoustics Research Department. We at least had a relatively easy time explaining how our research could ultimately lead to products that would be relevant to the company. Starting 3 years after the divestiture, the Baby Bells (the local telephone operating companies) were given progressively more authority to enter the telecommunications market and in 1996, they traded their local phone call monopoly for the right to compete directly with AT&T in the long-distance telecommunications market.

The consent decree of 1984 and the 1996 decision to permit the formerly local operating companies to compete in the long-distance market had a dramatic impact on AT&T and its Bell Labs. The corporation was now much smaller and needed to focus on competitiveness, an issue that had never been important before, as AT&T had thought of itself and largely had been a public utility. After the 1996 decision, AT&T management chose to break up the corporation into three pieces, an action that was known as trivestiture. The AT&T name would stay with the portion of the company focused on communication services and financial services through the AT&T Card. A second part of the business would focus on communications products and equipment. This new entity was named Lucent Technologies and Bell Labs was incorporated into it. The third corporate entity reverted to its earlier name of NCR (formerly National Cash Register) and focused on ATMs.

Although it took several years for the impacts to shake out, the role of Bell Labs changed considerably. No longer would the Labs be a stand-alone unit focused on research, funded by taxing other units of the corporation. Bell Labs would now need to be profitable (at least to some extent) on their own or at least able to show a direct link to a profitable product. It is hard to overstate the enormity of this change from a research perspective. Basic research tends to ask fundamental questions about how or why things are the way they are. Once answers are understood, it is conceivable but not guaranteed that an application of the knowledge will be found. This might include a new device, or a means of harvesting or controlling or monitoring whatever process has been studied, but such developments take long time periods to develop. A current example is offered by quantum computing. We have understood quantum mechanics for decades, but only now are we beginning to see how that knowledge might let us build new types of computers capable of performing computations that would otherwise be impossible. The decision to force Bell Labs to align with products would force the time horizon to move in from decades to months or years. It would stifle work on the most fundamental and unknown processes, precisely those that if harnessed could transform the industry.

Many of us inside Bell Labs wrote white papers about the damage that would be done to Bell Labs by insisting they become profitable, but we were being presented with the stark contrast between AT&T Bell Labs and Hewlett Packard's research unit. Hewlett Packard was earning orders of magnitude more on their intellectual property development than AT&T was, despite AT&T having many more important patents. Of course, this ignores the fact that AT&T put most of their research inventions into the public domain, including the transistor, the laser, and the electret microphone. Ultimately, the decision was made to push Bell Labs into a profitable stance. I would argue that this decision was the death knell for Bell Labs. It was simply unreasonable to assume that research focused on the long term would continue to be well supported by companies reporting to investors with a short-term focus.

Toward the end of my time at Bell Labs, I was asked to work with a team in Indianapolis put together to commercialize acoustic devices. I ended up managing a team of ten to fifteen people designing and manufacturing specialized microphones and speakers and I'll admit that parts of this task were fun. It was interesting to see how a design based on prior work in the research arm of the corporation could come together. We designed and made directional microphones for telephony, using a boot on a standard microphone that altered the delay between sound hitting the front sensing surface and hitting the back sensing surface. This boot let us set the delay at the optimum for the desired directional response. We made second order directional microphones and rabbit ears, loudspeakers that you could put on the two sides of the screen of a computer to deliver stereo sound. We also made a successful echo canceller, a device to prevent echoes when speakers and microphones are used in close proximity. That design eventually was used in the very popular Polycom trident telephone, a triangular-shaped conference telephone with microphones at its vertices and a speaker in the center.

This commercialization task required me to fly to Indianapolis every other week for mandatory meetings. Additionally, I would go to Indianapolis as needed. The point is that I put on lots of miles and time away from home in working with the commercialization team. It wasn't ever the sort of task I would have volunteered for, but Arun Netravali, the Bell Labs President at the time, made it clear that he thought the only way the labs would survive was if people like me agreed to help commercialize the research results. I felt obligated to give it a try since Bell Labs had been so good to me over the years.

The bottom line is that the team I helped guide had a healthy portfolio, had some talented staff, and was doing reasonably well for a startup until its leadership got greedy. Because of the pressure to show a profit quickly, the management of the commercial venture wanted to ramp up the profit margins on products exponentially. For instance, the electret microphones, complete with an amplifier attached, were initially being sold for something like 20 cents each, and cost about 5 cents each to make. The decision was made to charge $10 each for them instead. This cost meant that the microphone might become the most expensive component in a product, so it was a very unpopular change with customers. Combined with other rapid price hikes, the venture started to spiral and ultimately failed.

While I found it interesting to see products coming together, I am not temperamentally suited to commercial production. Once I understand how a device works, I have little interest in the details of its design and even less interest in squeezing the last penny out of its manufacturing. I am truly a researcher, and I was being asked to turn myself into a product designer. It clearly wasn't where I wanted to spend the rest of my working career.

The Indianapolis team I'd been working with opted to spin out of AT&T. They reasoned that as a startup they would potentially have a lower overhead and be able to reduce the costs of products without pressure from managers to account for a significant profit on a large corporation's bottom line. When the Indianapolis group formed that stand-alone company, I was asked to lead it. Much as I did when offered the chance to start a company building electret microphones, I turned the offer down because it wasn't to my liking. I'm more interested in what does not exist than in taking advantage of a bunch of things we understand. I had already known this about myself before this stint in Indianapolis. I had been offered a few opportunities to move into low level management positions at Bell Labs and had turned them down. They didn't fire my passions and get me interested in getting to work in the morning; nor did they provide a clear avenue for advancement as Bell Labs still had a lily-white management team.

I might have stuck it out with this new model for Bell Labs, as I still had my research lab and activities, but these too changed. The new model gave the profit-generating sections of the corporation the right to decide whether to support proposed research directions. Of course, these corporate entities wanted to support what they could see they needed, not basic science that might not produce a new product for decades. I still had questions I wanted to pursue, and I tried to shop around for support from within AT&T, but I couldn't find it. With much regret, it was time to move on.

Although I was of retirement age according to Social Security and many of my friends and family, I never once considered just leaving Bell Labs for the slow, comfortable life of retirement. I couldn't then, and still can't, imagine what I would do without an ability to ask technical questions and think about the answers, testing my theories in a lab. It would simply require an impossible personality change. Instead, I started to think about the opportunities to move into academia, the only remaining place where my kind of research would be appreciated.

I was not alone in thinking about a move to academia. The standard academic measures of success include the number of publications and patents published, the number of invited talks given, and the awards received. In these measures, Bell Labs research staff tended to look better than their academic counterparts, in part because they didn't need to spend time writing research grant proposals or in classroom teaching. Several of my friends and colleagues at the lab chose to make a move to academia rather than bear witness to the destruction of Bell Labs. Jim Flanagan, my department head and speech acoustics expert, ultimately left and became Vice President, Research at Rutgers University. Larry Rabiner, a speech signal processing leader, joined Jim Flanagan. Max Matthews, one of the fathers of computer-generated music and our center director, left for Stanford. Jont Allen, a researcher in hearing, went to the University of Illinois. Bill Massey, one of the originators of financial engineering, left for Princeton University. Bill Wilson, an expert on nanoscale systems, left for Harvard. Victor Lawrence, working on data signaling to increase transmission speeds, left for New Jersey Institute of Technology. It was as if the best and brightest of Bell Labs were dispersing throughout the country—a breaking up of the illuminati!

I visited five universities, ostensibly to give talks, but really to see if there was an appealing culture and group with which to collaborate. I went to Georgia Tech, Penn State, MIT, Stanford, and the University of Illinois. Although none were quite the same atmosphere as Bell Labs, I finally decided to pursue an appointment at Penn State, the only place in the world with a degree-granting department of acoustics. While the people there were warm and talented, and the presence of the largely Navy-funded Applied Physics Lab meant there would potentially be lots of supported opportunities for research collaboration, I was concerned that Penn State is in the middle of nowhere. That would make it harder to work with collaborators outside of Penn State and would limit the access to social amenities. Nonetheless, it looked like the best option for me, so I started the ball rolling.

Although I had a long history in the fields of acoustics and signal processing, the formalities of getting an appointment at any university meant I had to line up people to provide letters of reference for me. I decided to ask Ilene Busch-Vishniac, at the time the dean of engineering at Johns Hopkins University (JHU), if she would be willing to provide one of the references. I called Ilene and we had a nice conversation, briefly catching up. I filled her in on the problems at Bell Labs and my intention to move to Penn State. She commiserated with me over the demise of basic research at Bell Labs and immediately agreed to serve as a reference for me. About 10 min after we concluded our call, Ilene called back. She said that if I were serious about leaving Bell Labs, she'd like me to consider moving to JHU, although she didn't put it in those terms. She basically said, "What are we, chopped liver?" As she pointed out, Hopkins was a place that ate, slept, and breathed research. It was much more suitable for me than Penn State in her opinion.

Ilene arranged for me to visit the Electrical and Computer Engineering (ECE) Department at JHU and stepped aside so she wouldn't bias any decisions. I left with the impression that JHU was more like Bell Labs than any of the other places I had visited. Most of the other schools seemed to be stovepiped—collaboration was always within the group rather than across fields, groups, and departments. That type of culture set up a competition between the groups rather than the collaboration culture in which I thrived. At Hopkins, this stovepiping was not the norm. Virtually everyone in the ECE Department was collaborating outside of the department, typically with folks in Biomedical Engineering, the School of Medicine, the School of Arts and Sciences, the Applied Physics Labs, or Peabody School of Music. It was expected and rewarded. Collaboration barriers had been lowered through many policies in place, and I was sure I'd fit in with the culture. Hopkins also had the advantage of being in Baltimore, so it was urban with lots of cultural offerings. It was a city with a large Black population, one that spanned the financial ranks. It also had a lovely waterfront and was accessible to Washington, DC with just a short train or car ride. Although my previous experiences in Baltimore had not been entirely positive, I was sure that I could be happy there.

In 2001, I was offered the position of Research Professor in the ECE Department and accepted the offer immediately. Being a Research Professor meant that I would not be required to teach formal classes, and would need to support myself on research grants, but that I could supervise MS and PhD students and run a research lab. It was very close to continuing the situation I had at Bell Labs. I would be permitted to view the world with 360 degrees of vision and to collaborate with researchers anywhere. The only real shift would be in application focus—from communications for AT&T to medicine and health, the unstated but understood focus of JHU. Officially, I retired from Bell Labs.

The biggest problem with leaving New Jersey was my family. Melanie and Laurie were no longer living near us. Melanie was in Chicago and Laurie was in western New Jersey but Jay and Jackie and their family were still nearby. It would also make day trips by Laurie less likely since she was living about an hour and a half drive away but not in the direction of Baltimore. That would make it difficult to leave.

By this time, we had moved from 510 Parkside Road, the house with the recording studio in the basement, to 724 Berkeley Avenue. The move from an up-and-coming Black community to a predominantly white community was prompted by Marlene's desire to live in an area more like where she grew up. Ellington had gone to school in the predominantly white school district. As I was contemplating leaving, she was a rising junior in high school and had just been elected the president of the junior class. She was loathe to leave Plainfield and she couldn't live there alone. Marlene and I agreed that she and Ellington would remain in the Plainfield house while Ellington finished high school, and then Marlene would follow me to Baltimore. I would find a small place near the JHU campus to live in temporarily, with the aim of finding a new home for Marlene and me once she was able to move to Baltimore. In the interim, Marlene would continue working. I would commute via Amtrak from Baltimore to New Jersey on weekends. Every summer, we would rent a large house for the extended family to stay in on the New Jersey shore for a week or two.

Although Ellington was doing well socially in school, she was struggling academically at the time. On my weekends home, I tried to help her with mathematics, but Ellington complained that my explanations were going over her head. It was a source of frustration for both of us. Mathematics comes so easily to me it was painful to see Ellington struggle with it.

Ultimately, the two-year delay for Marlene's move expanded considerably. By the time Ellington was nearing an end to her high school days, Marlene's mother was nearing the end of her life. She had long suffered from rheumatoid arthritis and was now wheelchair bound and reliant on her husband for support. Unfortunately, my father-in-law was a cruel man, not well suited for the caregiver role. On occasion he would tie my mother-in-law into her wheelchair and put her in a closet. Generally, he neglected and mistreated her in these, her waning years.

Although Marlene was not the only child of my in-laws, the task of intervening between her mother and father fell to her. She became the primary caregiver of her mother and made regular visits. Complicating this interaction was the insistence of both her parents that they would manage on their own, without organized support from social agencies or assisted living. They also resisted any attempts that Marlene and I made to set them up in a more comfortable and caring situation. In the end, Marlene's mom died about a year and a half after Ellington left Plainfield to go to college at the University of Maryland, Baltimore County (UMBC). That meant that we had lived apart for between 3 and 4 years, and I had logged many, many miles commuting to visit Marlene.

At the point that I left New Jersey, Marlene was teaching special needs kids in elementary school. When her mother's arthritis became crippling, Marlene quit working as a teacher to take care of her mother. She had earlier gotten her real estate license and opened a real estate firm with a partner and she returned to real estate at this time. It made her working hours more flexible and allowed her to care for her mother. When Marlene's mom passed away, Marlene was significantly depressed and pretty much unable to work for a couple of years. At that point she was home alone because I was in Baltimore and Ellington was in college. As she overcame her depression, she resumed teaching special needs children.

Like all married couples, Marlene and I had our differences, but we also had our points of agreement. We are both liberal politically. We both showered Ellington with love and support. We certainly agree on the importance of education above all else.

Our points of tension come from differences that are major components of our makeup. I am a scientist. Marlene doesn't trust science. She is against vaccines, for example. She would show me articles purporting to support her position, but I knew the "science" behind them was flawed. She didn't want to hear it. Marlene is also against, and distrustful, of big business. As a Bell

Labs employee, I owed my career to big business. I would not have flourished outside of the research facilities that a large corporation could offer.

I also had trouble understanding Marlene's family. When Marlene's mom got ill, there was a medication she should have been taking but refused because of the cost. I offered to put Marlene's parents on my Bell Labs insurance, but they wouldn't permit it. It would have required them to gift their house to someone else in the family so I could declare them dependents, but their living situation wouldn't change. I think their pride would not allow them to do that. The net result was that Marlene's mom suffered more than she needed to. It would not have happened in my family. We would have banded together and taken whatever action was needed to get the best treatment.

I understand that some marriages can survive separation and thrive on long-distance relationships, but I am not well suited to such relationships. I go where the wind takes me, and long-distance relationships require much more planning and cooperation than cohabiting relationships for them to be a success. I remember being asked by a colleague at Echophon, a sound absorbing material manufacturer, what it was like to have a marriage with a geographical separation. I told her that it was more like having a mistress than a marriage. You get used to living on your own and not consulting on important decisions. I was commuting to Plainfield from Baltimore on most weekends but finding this increasingly difficult.

The years of separation drove a wedge between Marlene and me, one that took decades to heal. Each of us had gotten used to being on our own and to our respective lives in Plainfield and Baltimore. Marlene was willing to move to Baltimore but only if she could be assured that I would once again be the loving and committed partner she had earlier. She suspected that I had been seeing other women while we were separated geographically and wondered if I was ready to stop that. Was I prepared to once again have a committed monogamous relationship with her? I was no longer sure I could offer Marlene that, and I told her so.

Truthfully, I am more cut out to be a recluse than I am to have a close relationship with anyone else. I hear people say, "I'm not complete unless I'm with the person I'm living with," but I've never felt that way about anybody. I've always felt completely independent. Most of the social part of life is only important to me if it involves something around a problem that I am trying to solve. Despite this sense, I enjoy putting people together who have something in common. I can see the common traits and see where the match is going to work as well as where it will be challenging. Friendships are important but not necessary.

At the time, I honestly resented Marlene not living up to her part of the bargain of moving to Baltimore when Ellington was out of the house. She was living a nice life in Plainfield, in a large house in a nice neighborhood while I was in a tiny one-bedroom apartment in Baltimore. It didn't seem fair to me. She had been faced with a choice between her parents and me, and she chose her parents.

Despite the souring of my marriage, I was happy at Johns Hopkins University and was learning the ins and outs of being a faculty member.

17

I Work at Johns Hopkins University

I joined JHU as a Research Professor in the Electrical and Computer Engineering (ECE) Department in November of 2002. At the point I joined JHU, Ilene was still the Dean of Engineering, so I had a very busy but local collaborator. Ilene stepped out of the dean position in June of 2003 and resumed research and teaching responsibilities. At that point we had much greater interaction.

One of the things I quickly learned about academic life is that faculty are asked to do much more than research. Typical faculty on the tenure track teach, serve the institution in some manner, and conduct research. As a Research Professor, I was theoretically immune from everything except research, but that theory didn't work out in my case. At the point I joined the ECE Department, I doubled the number of Black faculty members they had, so I got noticed. Department Chair, Gerard Meyer, came by every now and then to pick my brain on what we might do to increase the racial diversity of our student body. Gerard noted that we had a large HBCU with an engineering program almost in walking distance from our campus, yet we attracted few students from Morgan State University. Ultimately, Gerard offered to pay for shuttle buses to connect the Homewood campus to Morgan State's campus, so that Morgan students could take classes at Hopkins that weren't offered at Morgan. Unfortunately, with a long history of underfunding of HBCUs in Maryland, Morgan State University didn't buy into the program Gerard was proposing, instead getting defensive about their students and programs. To this day, the opportunity for strong collaboration between Johns Hopkins and Morgan State has been missed.

J. E. West, I. J. Busch-Vishniac, *Mic Drop*, https://doi.org/10.1007/978-3-032-20922-1_17

At the College level, I served on the Academic Council and as Chair of the Diversity Council from 2004—2007. I was simultaneously on the University Diversity Council, serving there from 2004—2008. These bodies advocated for greater diversity among the faculty, staff, and students. I knew what the reaction would be when we presented greater diversity as a request. It would match what we were told at Bell Labs: “You find them, and we will educate/hire them.” My reaction also paralleled what we found as a solution at Bell Labs—we needed to create the students and faculty to hire. I was among those who met with JHU President Ron Daniels to present our findings and requests. This led to the rebranding and expansion of the Baltimore Scholars program that had promised students in the city of Baltimore a generous scholarship if they qualified for entry to Johns Hopkins. The program, renamed the Cummings Scholars in memory of Representative Elijah E. Cummings, includes students from Baltimore City and Washington, DC. public and charter schools. It offers a free ride for students whose families earn less than $80,000 per year and caps the total expense at 10% of income for families earning between $80,000 and $150,000 per year.

Ilene enjoyed teaching an introductory course in acoustics and I helped her make it more hands-on. I remember we set up a way for students to turn on their music through headphones or earbuds, and to see just how loudly they were listening to the sound. I hope this encouraged the students to turn the volume down. Our students wandered the campus measuring common sounds. We also had them doing a rough calculation of reverberation times by dropping a book in a classroom and recording the sound on a first generation, handheld, digital recorder. The course was popular, and we had good fun teaching it.

Upon arriving at JHU, I looked around for opportunities for research collaboration. I spoke with faculty in my department, some of whom were working in areas tangentially related to acoustics, such as speech, and I reached out to the Applied Physics Lab, a large, navy-dominated, research unit of JHU. In relatively short order, I had students working with me.

I found myself with seven students relatively quickly—four undergraduates and three graduate students. The undergraduates were Bobby Ng, Douglas Orellano, Tyrone Hunter, and Ram Chivukula. My first graduate student at JHU was Dawnielle Farrar, who worked at APL. Colin Barnhill and Josh Atkins quickly followed as the second and third graduate students. The undergraduates had largely found me through a talk I gave in the department. Bobby was the exception. He was leading the go kart racing team at JHU, and I had naturally gravitated in that direction to see how things were going. We got to chatting and he took me for a ride. He was even kind and trusting enough to let me drive. Colin had also found me through my talk and wanted to work with me toward his PhD. Colin came from an interesting background. His parents were both lawyers and were managing the law firm Michelle Obama was in. Colin offered to have his parents introduce me to the Obamas, but I felt obligated to decline. I didn't want anyone to think Colin's progress was due to my gratitude for non-work-related meetings he had set up.

For me, mentoring students was challenging. We all come with personal problems and likes and dislikes but having been fortunate enough to make a substantial contribution to the advancement of acoustics, students put me on a pedestal. I spent a lot of time tearing that pedestal down. I'd tell students I'm just an ordinary human being who didn't know more than you when I started out. The main thing that I try to transmit is that there's nothing special about me and yet I have been very successful. You must want to be successful and put your mind and energy into making that dream come true.

Particularly for students of color, I talk about the 40 or so Black inductees into the National Inventors Hall of Fame. I choose these scientists because they have invented something that is important in the context of everyone's life. I describe what these people have done, and the conditions under which they contributed. I ask students to look at themselves, see what advantages they have compared to these inventors, and ask themselves what they would be happy doing the rest of their lives. The question I get back when I take that approach is, well why do you think I can do this? It is followed by the comment that you're giving me a real tough job, one that I'm not convinced that I can do myself. That's when I explain that failure is not only acceptable, but an essential part of the process. Most people try to ignore failure, but I know that some of the "aha" moments come from failure. The important point is to understand what caused the failure, what it is telling us about the nature of the problem we are investigating, and how to avoid making the same mistake in the future.

Dawnielle Farrar (now Dawnielle Farrar-Gaines) was my first Black graduate student and working with her posed some interesting logistical problems. While APL encouraged employees to pursue a non-thesis MS degree as a part-time student, they expected the courses to be from those offered at APL at lunchtime, early in the morning, and in the evening. No release time was provided for pursuing such a degree. At first, APL wanted Dawnielle to continue to work full time at APL while being a PhD student. Fortunately, the Director of APL, Richard Roca, had come from Bell Labs and he understood the problem when I explained that this wouldn't work for a student in a dissertation program. I convinced APL to grant Dawnielle roughly half time free from her normal APL commitments so that she could pursue her research. I wish I could say that this became a standing program at APL, but I don't believe that is the case. Dawnielle was a one-off.

Dawnielle's research work was on a particular polymer, PBLG (poly-y-benzyl-L-glutamate). PBLG was known to be a polar material, as it was commonly used in electrical switches. Dawnielle created fibers of PBLG in a drawing process and showed that these were also polar materials. She showed that they could be made piezoelectric, meaning that straining them produced an electrical signal. Upon graduating, Dawnielle continued at APL where she became a senior electrical and materials engineer working on systems involving micro and nano materials, extreme materials, microelectronic integration, and a variety of sensing methods. She also is on the faculty of the Material Science and Engineering Department, where she teaches a graduate course. Not surprisingly, Dawnielle has followed in my path in terms of actively encouraging minority students to pursue technical education.

We grew into a strong research team. The students helped calibrate and put together the equipment for a new laboratory in the space provided, including a reverberation/anechoic chamber. This chamber was convertible by removing absorptive panels and the heavy jute rug we had had made for the floor. We used the chamber to measure the sound absorption of various materials (in reverberation mode) and to study directional devices (in the anechoic mode).

One of the first uses of the lab came about serendipitously. We got a request for help from the Chief Information Officer of JHU, Stephanie Reel. She had adopted the Pediatric Intensive Care Unit (the PICU) at Johns Hopkins Hospital and would round with them monthly. There was generally not much she could offer as assistance, but on one occasion, the staff were explaining that communication was terribly difficult because of the L-shape of the ward and the equipment noise. Stephanie told them she knew just who to contact and got on the phone to Ilene. We set up a tour of the PICU with the nurse manager, Claire Beers, and assured them we would help them find a solution. We thought it very likely that there was a significant amount of work already done on hospital noise and communication, but when we looked at the literature, we found almost nothing relevant. Colin took this on as his research project and led the work we did to address the hospital noise problems.

The primary issue for the PICU was improving the ability for staff to talk with one another without having to leave the bedside of a patient. We found the ideal device for them—the Vocera communication product and recommended it. The Vocera badge (sized like a very small mobile phone) allowed for hands-free person-to-person communication using the telephone network. Staff wore it on a lanyard around their neck or pinned to their uniform. It responded to voice commands and was ideal for connecting people without requiring them to press any buttons. It eliminated much of the overhead paging and was considered a good solution by the staff, who quickly became experts on its use.

While we were working with the PICU, we wandered around the hospital and found it to be quite noisy. We managed to meet with the President of the Hospital, Ron Peterson, to suggest that we do a small-scale study. We asked for very modest funding to support a couple of students and the purchase of a sound level meter and materials. Fortunately, Ron had recently had a stay in the hospital and complained that he was unable to sleep because of the noise. He said yes to our project and even agreed immediately that we would be able to publish what we found. Armed with that commitment, we set to work measuring the sound pressure levels (an objective measure of the sound intensity on a logarithmic scale to mimic human perception) at various places in Johns Hopkins Hospital. We made measurements of daytime and nighttime noise, finding that they were nearly the same level. We wrote an article documenting what we found. This first of four articles on hospital noise became one of my most cited papers. When Ilene and I spoke about our work, we were surrounded by acoustical consultants who wanted to know how we had gotten permission to publish the work since they had been prohibited from publicly disclosing most of their work measuring noise in public institutions. It confirmed our confidence in JHU that they understood how important it was to publish research, even if it made the hospital air some dirty laundry.

With one acoustics problem at the hospital solved, we were asked to work with another unit to see if we couldn't find a means of quieting the din. Claire had recommended us to Anita Reedy, the nurse manager of a blood cancer unit in the Weinberg building. We went to examine the unit and found perhaps the worst possible architectural design. The rectangular unit had not the normal perpendicular corners, but a small 45-degree connector between two adjoining walls. That meant that sound could run down a hallway and be reflected into the perpendicular hallway. It could literally loop around the unit. We found that conversations anywhere in the unit could be heard pretty much everywhere. Further, the issue of keeping the unit clean meant there were no traditional sound absorbing materials anywhere—hard walls with no drop ceiling and absorbing tiles. The unit also had a weird, large circular "feature" in the ceiling at the central nurses' station that tended to trap the sound.

We knew we couldn't use the standard sound absorbing tiles in the hospital because they can't be cleaned. We decided to see how much their performance would degrade if put in a plastic bag. Our lab reverberation chamber was used to determine the impact of placing these tiles in a normal garbage bag, and a contractor's garbage bag (which is three times thicker). We found that the sound absorbing performance degraded some but not greatly with enclosure in these plastic bags. We found a material that was rated for use in hospitals and that could replace the garbage bags. Thus, we took the next step and installed a series of sound tiles in this new fabric high on the walls around the unit, with emphasis on the reflecting walls in the corners and the interior of the odd circular feature. As soon as these went up, the sound pressure level on the unit perceptibly dropped and the phone ring tones were turned down. This experiment proved we could make the unit quieter. This work was published as another of our series of four on hospital noise.

The remaining two hospital noise articles were on the noise in the Emergency Department, and the noise in the operating rooms. The latter of these two was quite interesting because it meant that we were given access to surgeries to monitor the sound, once again showing the trust Johns Hopkins put in us. Unsurprisingly, we found that orthopedic surgery tended to be very noisy, but none of the surgeries were particularly quiet. This work became the doctoral dissertation for Colin Barnhill. Colin joined the JHU Institute for Policy Studies upon graduation and then became an acoustical consultant for Johns Hopkins Medicine.

We concluded our work on noise at Johns Hopkins Hospital with a conference talk on the work of one of our undergraduate students, Phil Kwon, on speech recognition in the hospital, i.e. on the ability of listeners to understand what was said to them. Our measurements showed that speech communication in the hospital was degraded significantly by noise. At best, the sound environment was shown to produce mediocre speech communication for normal hearing individuals. This was determined using standard measures of the sound pressure level of the background noise to produce a speech intelligibility index, a standard technique used by acousticians. Of course, hospital patients should not be considered normal hearing individuals as they are likely medicated (which affects concentration and possibly hearing) and elderly, suffering from normal loss of hearing acuity with age. Thus, the study probably overestimated the quality of speech communication in the hospital.

Our speech communication study was important because much of the communication to patients in hospitals is oral rather than written. Although we never published this work, it was later supported by a comprehensive study done by Erica Ryherd and her students at the University of Nebraska-Lincoln. Erica's work showed that speech communication in hospitals is, at best, mediocre.

For the most part, both Ilene and I stopped working on hospital noise after this set of papers. We share a common approach to research. Once we understand the science, we tend to lose interest and want to move on to another topic. This was certainly the case for the hospital noise work. We had learned the major outlines of what was going on and moving into the finer points, such as designing new equipment or ward architectures, didn't really appeal to us.

Simultaneous to our work on hospital noise, Josh Atkins did a nice piece of work under my supervision on sound localization while wearing headphones. If you have headphones on and rotate your head, the image you have of the sound source location rotates with you. What Josh was able to do was use sensors on the head to determine its orientation and then stabilize the sound source image using real time processing. In this system, turning your head simulates the real experience of being in an acoustical environment and turning your head. Upon completion, Josh first joined Beats Electronics as a founding member, then joined Apple a few years later as they established an acoustics department. As of 2025, he is the Senior Manager for Audio Algorithms, leading a team focused on machine learning for speech and audio applications, spatial audio algorithms, and multichannel signal processing for dereverberation, noise reduction, and blind source separation.

Shortly after arriving at Hopkins, I was invited by Dr. Ben Carson to speak with his student group. At the time, Ben was still working as a neurosurgeon, and he was also very active in a program he and his wife crafted to identify exceptional students and groom them for college. Ben wanted me to participate in his program, but I declined for a very selfish reason. Having grown up in the south and suffered the effects of white privilege, I was determined that my advocacy work would focus on opening opportunities through programs that were majority Black students. Ben Carson's program didn't meet that criterion.

Trying to persuade me to participate, Ben invited my family to his home for dinner. There were a couple things over the course of the evening that struck my family and me as quite odd. There was a large painting over the mantel in the living room that portrayed Ben with Jesus, side by side. While I am not a religious man, this seemed blasphemous and to be a display of amazing self-regard. When we went to dinner, I noted that the children were separated from the adults and dined elsewhere. That is not how I was raised. Meals were a family gathering and adult conversation was meant to be educational for the children. It was a pleasant but awkward evening.

This was not my last encounter with Ben Carson. Some years later, when friends organized a conference in celebration of my 80th birthday, I invited Ben to attend and speak so that he could see how I viewed racial issues and had made progress. His talk, though, was deeply religious and not scientific. It was not appropriate for the forum. Fortunately, or unfortunately, depending on your political perspective, Ben left medicine and went into politics, ultimately serving as the Secretary of Housing and Urban Development under Donald Trump. His calendar was sufficiently busy that we never bumped into one another again.

Over the years, I've heard many excuses for why educational programs in the sciences tend to be so dominated by white males. Chief among these excuses is the pipeline argument—that the interest of women and people of color in the sciences wanes as they go through school, making it nearly impossible to find candidates for advanced scientific programs from these groups. While I would argue that traditional educational programs are structured to produce this result, what ABLE and CRFP showed me at Bell Labs is that there is some truth to this argument, although it doesn't describe the challenge fully or excuse entrenched programs that don't ask how to affect changing this situation. The Bell Labs solution was to create programs that reached further down the educational pipeline, to create students from underrepresented groups who would be able to participate in CRFP (and the equivalent for women) once they graduated with their undergraduate degree in a STEM field. It was this thought that created the Summer Research Program for undergraduates.

With the educational pipeline in mind, I can look back now and be very proud of having participated in programs that aim to encourage women and underrepresented minority students to move into STEM careers along the entire pipeline length, from kindergarten through high school and college. While ABLE focused on the college and university end of the pipeline, I have been able to focus my program involvement on the grade school end of the pipeline in my time at JHU. I have done this through engaging with Camp Invention, the summer camp program of the National Inventors Hall of Fame, and with the Ingenuity Project in the Baltimore City schools.

Camp Invention was established in 1990 and is a weeklong STEM summer camp for kids entering grades K – 6. It brings curious kids together and encourages them to construct creative inventions and form long-lasting friendships. It builds confidence, teaches problem-solving skills, reduces summer learning loss, and improves academic achievement. It also introduces campers to role models, often inductees into the National Inventors Hall of Fame, who can share inspiring stories and encourage these campers to follow their curiosity where it leads them. I am shown surrounded by a group of Camp Invention students in Fig. 17.1. Founded in 1987, Camp Invention has more than 1000 locations nationwide, including at least one in every state and in Puerto Rico. More than 122,000 students participate every year in Camp Invention programs. A program for middle school children entering

Fig. 17.1 Camp Invention gathering with me

grades 7–9 called Leader-in-Training has been added and is a more advanced version of Camp Invention.

The 2025 topics for Camp Invention are Illusion Workshop (video special effects), Claw Arcade (building a claw machine), Penguin Launch (robotic rover), and In Control (car control panel). In each case, students build their own inventions, gaining valuable hands-on experiences and producing successes to encourage further creativity.

After being inducted into the National Inventors Hall of Fame, I was asked to get involved with Camp Invention. I found that the program was overwhelmingly attracting white children from affluent families. It was easy to see why. Liberal arts sorts of programs are relatively inexpensive to run, but science programs require equipment and greater supervision to prevent accidents. This means science programs tend to cost more. At the time I joined the board of Camp Invention, the cost was being passed along to families as a set fee, and it was primarily the affluent white population that could easily afford this investment. I convinced the board to change their funding model to have a greater Robin Hood component, i.e., to raise the cost of the program for those who could afford it, and to offer more camp scholarships for families that could ill afford the camp costs. I also insisted that more of the camps be in inner cities so that residents in those areas could easily get to them. These actions have dramatically altered the demographics of Camp Invention, which now serves a very diverse population.

I would have loved participating in Camp Invention had it existed when I was growing up. When I first saw the program, its facilities consisted more or less of a room full of spare parts from electrical and mechanical items that were taken apart. Campers were encouraged to try swapping out one component for another to get the behavior they wanted. This would have made me a very, very happy camper. The program is now much more structured, not only to prevent safety hazards, but to make sure that campers will be successful in their invention builds. It too would have won my heart.

The Ingenuity Project in Baltimore was founded in 1992 at two middle schools and funded by the Abell Foundation as a means of responding to the desire to provide better science, technology, engineering, and math learning opportunities for students. The program grew and in 1997, it was expanded to include programs for high school students. The Ingenuity Project still lists their mission as preparing and launching "the next diverse generation of nationally competitive STEM leaders from Baltimore City Public Schools." The Ingenuity Project is in four middle schools and the science magnet high school in Baltimore. The catchment area includes 29 zip codes, and more than 800 students are participating.

The Ingenuity Project middle school program funds extraordinary math teachers and provides a rigorous curriculum so qualified students can learn in classes of no more than 15 students. The courses emphasize inquiry, investigation, analysis, and strong work habits. The high school portion of the Ingenuity Project has two tracks: a research practicum and an innovation practicum. The research practicum is for qualified students entering grades 10–12 and has students work with mentors in local colleges, universities, and research labs. Students perform mentored independent research and submit their work to national pre-college competitions. The innovation practicum is for students in grades 10–11 and has students working with local companies and universities to gain hands-on experience in the application of science in products. Nearly 100% of the Ingenuity Project students apply to and are admitted to competitive university programs such as MIT, Harvard, Yale, and Johns Hopkins University. All students are given personalized guidance for post-secondary options. Most of the students have scholarships to cover the full cost of the college years.

Shortly after I arrived at Hopkins, I was approached by Professor Jeff Gray in the Whiting School of Engineering. He had been serving on the Ingenuity Project board and was concerned that the program was predominantly white and male in a city that is roughly two-thirds Black in its public schools. Jeff thought his concerns would be better expressed by me, and hoped I'd be willing to take over the board seat.

Jeff was, of course, offering me an opportunity I couldn't decline, so I joined the Project Ingenuity board and began to investigate why the program was not more representative of the Baltimore City population. What I found didn't particularly surprise me. The Ingenuity program had an application process from which it chose participants. The application asked for a recommendation from a teacher. It was these recommendations that were holding the Black students back. Two students with identical credentials might apply, one white and one Black, and the Black student's recommendation would be mediocre while the praises of the white student were sung. Further, the typical bias of assuming Black children would perform poorly and their participation would require lowering the expectations was present. I asked the board to eliminate the recommendation from the qualification process, and they did. Now the program is about three quarters minority students and more than half female. Performance has not slipped one iota from what it was when applicants to the program had to get that recommendation.

I am a strong believer that change needs to be self-perpetuating. If I create a program that depends on my presence and then I leave, the whole effort could be an exercise in futility. For this reason, I am very pleased that the Ingenuity Project makes it a point to put some of its graduates on the board. These former students know precisely what the program has meant to them and will make certain that it is an opportunity available to future generations.

What the educational programs I've participated in and grown have in common is a focus on presenting technical problems and asking students to find a solution. I believe there is a cadre of people that, given a problem they care about, will come up with a solution regardless of their training, their background, or the tools available to them. Finding these students and providing them with the tools to grease their success is what I have tried to do in the educational programs I've created and worked with. Unfortunately, our public education system is lecture-style based on facts that are seemingly disconnected rather than problem-based. This partly explains why we don't achieve our national goals in the sciences.

While the US still holds a lead in some of the science fields, other nations have been catching up to us quickly, and I believe could surpass us if we don't change our public education style. The fact that we are not mortified by a pipeline that shuttles women and students of color away from the science fields dooms us, as we are removing well over half the population from participating in the most lucrative and important fields of endeavor. We need to get beyond misogyny and racism and change the way we think about teaching in the sciences so that we encourage everyone to be curious, to experiment, and to appreciate science.

In addition to these programs, I have long taken advantage of the opportunity to talk to young people. To be involved with technology is a rarity for them, because there are so few Black people in technical fields. Black people represent less than 5% of the STEM workforce. This empirically reduces the number of potential multipath collisions. In other words, the probability of a Black kid in the city of Baltimore meeting someone who's involved in engineering, in any area of technology, becomes very slim. For me, it is about exposure to the whole area of engineering, chemistry, mathematics, and sciences that may interest these kids. Knowing that there is a path for them heightens their desire to continue in the field.

The years 2000–2010 were the second decade in which I received a significant number of prestigious awards. I know that the best way to win an award is to win an award. In other words, once you get chosen for recognition, others tend to notice and want to join the party. Overall, I like being recognized, but I am not willing to focus on producing the mountain of documentation needed for some nominations. While I want to serve as a role model, I don't want to waste my time actively seeking awards.

In 2002 I delivered the Audio Engineering Society Distinguished Richard C. Heyser Memorial Lecture. My talk was titled *Modern Electret Microphones and Their Applications.* The Audio Engineering Society was founded in 1948 and is an international professional organization for audio engineers, artists, scientists, and students of audio. The Distinguished Richard C. Heyser Memorial Lecture was established in May 1999 and aims to bring eminent scholars to the AES.

In 2005, I was named the Black Engineer of the Year from NSBE. This award was created to highlight Black engineers who have made outstanding contributions and is presented at an annual conference. I also received an honorary doctorate from Princeton University in 2005. This was followed the next year with an honorary doctorate presented by Michigan State University. In 2007, I received another honorary doctorate, this time from the New Jersey Institute of Technology. At some point, I also received an honorary degree from the University of Pennsylvania. At that commencement, Vice President Biden gave the address. I had a minute or two to engage him in conversation and I pressed him on the need to improve opportunities for people of color. I'm not at all sure he heard what I was saying.

In 2006, I was presented with the Gold Medal of the Acoustical Society of America. The Gold Medal is the highest award conferred by the Acoustical Society and it is presented once a year. My citation was "for development of polymer electret transducers, and for leadership in acoustics and the Society."

Receiving a 2006 National Medal of Technology and Innovation from President George W. Bush was quite meaningful for me. This Medal is considered the U.S. equivalent of the Nobel Prize. It is the highest honor awarded for technology in the US. My citation read "For co-inventing the electret microphone in 1962, 90% of the two billion microphones produced annually and used in everyday items such as telephones, hearing aids, camcorders, and multimedia computers employ electret technology." I couldn't share the award with Gerhard because he was not a US citizen (Fig. 17.2).

It was great fun to receive the National Medal of Technology. It was presented in person by President George W. Bush. As this particular president was not well known for his scientific abilities or his respect for science, I was teased by my colleagues about having to spend time with him. There was a lavish dinner thrown after the award ceremony attended by many of my Bell Labs friends and supporters, including former boss Jim Flanagan.

In 2008, I delivered the Dr. Henry Hill Lecture of the National Organization for Professional Advancement of Black Chemists and Chemical Engineering (NOBCChE). The aim of NOBCChE is to promote a more diverse community in STEM fields and to advance the careers of Black chemists and chemical engineers by supporting their careers. The Henry Hill Lecture recognizes a Black scientist whose career can serve as an inspiration to others. That same year I also gave the Harold W. Gegenheimer lecture at Georgia Institute of Technology. For this lecture I spoke about decreasing the noise in hospitals.

In 2009, I received the Achievement Award of the Black Faculty and Staff Association of Johns Hopkins University (BSFA). The BSFA was established in 1995 and is the oldest and largest affinity organization at Johns Hopkins. Its goal is to promote the equitable treatment of Black people at JHU, and to serve as a resource for inclusivity and justice.

Fig. 17.2 I am presented with the National Medal of Technology and Innovation by President George Bush

In 2010, I received two prestigious awards: The Benjamin Franklin Medal of the Franklin Institute, and the Achievement Award of the National Technical Association. The National Technical Association had previously presented me with honors. I appreciated their recognition of my activity in encouraging Black children to pursue careers in science, math, and engineering.

The Franklin Institute was founded in 1824 in Philadelphia and offered its first awards in 1826. This makes it the oldest broad science and engineering award program in the country. The list of Benjamin Franklin Medal recipients is like a Who's Who in science, and includes Marie Curie, Albert Einstein, Nikola Tesla, Thomas Edison, Orville Wright, Jacques Cousteau, Jane Goodall, Bill Gates, and Frances Arnold. To date, 125 of the Franklin Medal laureates have received Nobel Prizes, most after receiving the Franklin Medal. I have no expectations of receiving a Nobel Prize, but being a Benjamin Franklin Medal recipient was a high point in my career. It meant I had come full circle, from reading about Franklin as a child and wanting to be like him, to receiving a medal crafted in his honor meant to go to those who were indeed like him.

The year I received the Benjamin Franklin Medal, Bill Gates also won a Benjamin Franklin Medal. I met him at the dinner for laureates and found him to be a quiet, thoughtful man. At that point, Bill was touting meat made from laboratory cell growth as a means of reducing our carbon footprint. While he was ahead of the technology, Bill's vision is now becoming a reality.

The decade of 2000-2010 clearly was a time in which I received many prestigious awards. It left me with a feeling of insecurity. I wondered why I was receiving all these awards for simply doing my job. Having grown up in a family where rewards were a rarity, it took me time to become comfortable with being singled out and recognized. Among the many awards, it is the Benjamin Franklin Medal which means the most to me. It connects to my youth, when I read about Franklin and determined I would be like him. It made me feel as though my goals had been obtained, my life made complete.

18

Medical Acoustics

When I first joined Johns Hopkins, I opted to live near the university for convenience. Marlene was going to join me after Ellington finished her high school years, so I thought I'd wait for her before making an investment in a house or condominium. I lived in small apartments, either studio or one bedroom, in an area generally dominated by students. While the apartments were small, there were many restaurants in the area and the convenience outweighed the noise and small size.

As the years dragged on without Marlene joining me, I opted to move to a nicer location. In 2005, I moved to Baltimore's harbor, renting an apartment on Thames Street. It did me a world of good to be back along the water again, in a vibrant area of the downtown. I spent more than a decade in that location and never tired of the proximity to the water, good food, and good music. The university was a 20–25 min drive away, but it was worth removing myself from the campus to find peace and tranquility.

Starting in 2015, I also had a roommate. Louisa's brother, Henry Moss has a son, Ian, my godson. Upon finishing his undergraduate degree and his service in the Marine Corp, Ian wanted to go to law school. The University of Miami offered him a scholarship that would have paid virtually all the bills, but Ian wanted to stay closer to his roots and attend school in Washington DC. I told him that it would be a long commute from Baltimore, but that if he were willing to spend the commuting time, he could live with me and avoid charges for room and board. Ian took me up on the offer. When he finished his law degree, he joined the Obama administration working on legal issues associated with the prisoners the US is holding in Guantanamo Bay.

J. E. West, I. J. Busch-Vishniac, *Mic Drop*, https://doi.org/10.1007/978-3-032-20922-1_18

This work continued through the Biden administration, after which Ian joined a DC law firm.

When Ellington finished her undergraduate degree at UMBC, she chose to stay in the Baltimore area. Ultimately, she and Josh Barnes became a couple and married in 2020. Early on they decided to buy a place together and asked me to join them. In 2017 we moved to a newly built condominium on the water slightly further east than I had been living. Josh's parents moved to the Boston Street development before we bought in, and having everyone around greatly increased my social interaction. We continue to live together there. The condo is very vertical, and it is right on the pier with balconies overlooking the water. I live on the second floor and Josh and Ellington are on the third floor. The first floor is the common area—kitchen, living room and dining room.

I have lived much of my life surrounded by an extended family, so this living situation, with my children and my son-in-law's parents in the neighborhood, feels natural and comforting. When we all moved in together, I still was fairly mobile, and I took pleasure in helping out with cooking. I was driving to campus 5 days a week and participating a bit more in the departmental operations, as my title had changed to tenured full professor in 2015. I made it a point to give Ellington and Josh lots of space to develop their relationship independent of me. As time progressed and my mobility waned, I became more of a spectator than a participant in chores, but I still found great pleasure in being ensconced within my family. Certainly, the cats and the dog have been happy to have me around more.

I have lived an active life from birth, participating in sports and generally moving rather than being a couch potato. But my body conspired to make this difficult, and ultimately impossible, as I aged. In the 1980s I developed some back pain from deteriorating discs, but I was content to live with this pain. Starting around 1998 I developed a footfall problem in one leg. The muscles controlling the lifting and lowering of the foot were weakening, leaving me with poor control of the foot dropping and raising motion. Having been trained to ignore my back issues, I ignored the footfall problem as well. When I finally went to the doctor, I was told that footfall problems require immediate surgery or they tend to become permanent, as mine had. The combination of the disc deterioration and the footfall problem affected my balance and thus my mobility.

By 2015 I started using a walker of sorts to steady myself as I moved. I began with a triangular walker that gave me handles to hold as I walked and that was fairly unobtrusive. This progressed as my orthopedic issues degraded my mobility, to the use of a rollator to provide more support. In essence, as I aged and my mobility problems increased, I was using my arms to support my weight more than my legs. My walk became progressively slower and more painful.

I pushed myself to continue to move for as long as I could. I went through physical therapy multiple times and had even forced myself to live in the Thames St. apartment, which meant I'd need to go up and down stairs, so that I wouldn't lose the ability to do stairs. After moving to Boston Street, we had a chair lift installed to run along the stairway so that I no longer had to manage going up and down these particularly steep stairs. Despite my work to retain mobility, my body has failed me in my waning years. I pinched a nerve in my back in 2022 that has made it impossible to use my arms to support myself. I opted for surgery and a stay in a rehab facility to improve my situation, but I am largely at the mercy of caregivers now to get around. Every move takes lots of preparation and focus.

I don't want to overstate the importance of my mobility issues. I am lucky to be able to afford accommodations that make it possible for me to continue to work and to live where I wish. Johns Hopkins continues to keep me employed even though I now interact with students remotely from home rather than an office.

I understand that the ravages of age tend to come in various forms. For many, dementia sets in and clouds the mind. I don't seem to be suffering from that. For most, physical issues arise that affect mobility. For the unlucky, both dementia and physical issues set in.

A positive outcome of my decline has been Marlene's move to Baltimore. After years of living apart, she finally retired and moved to Baltimore. Although we are not living in the same space, she has become one of my primary caregivers and we are more like a couple than we have been for a long time. Caring for me also provided the incentive she needed to overcome her alcoholism problem, although like all alcoholics, she occasionally reverts to her old ways.

Marlene is not my only caregiver. I now have professional caregivers with me 5 days a week. They help me with the various problems my lack of mobility has created—normal cleaning and dressing chores, for instance. On weekends, I rely on my family to help me.

Despite my mobility issues, I continue to be productive and to love the research challenges. I've found it hard to leave the university and when I tried initially, they asked me not to retire. We have finally gotten beyond that roadblock.

One of the great advantages of being at a university or a research company is that there tend to be regular talks about research, both from employees of that organization and from visitors. I have found on more than one occasion that these presentations have sparked a research avenue for me.

I know that some researchers have well-developed strategic research plans, but that doesn't describe me. I would say that my work life, like my personal life, has been a random walk. I find an interesting problem and work on it until I think I have an answer. Then I find another problem that sparks my interest and investigate it. This has given me the advantage of a broad repertoire of areas in which I've conducted research, and that lets me bring the knowledge of one field into application in another.

At one point in the early 2010s, I saw that there was going to be a talk at the Bloomberg School of Public Health on diagnosing pneumonia from stethoscope sounds. This seemed interesting to me, so I wandered over to the east campus for the talk. I learned that Drs. Jim Tielsch and Joanne Katz, public health researchers, were trying to tackle the problem of pneumonia in children, and that it is the top killer of kids under the age of five worldwide. In many of the places where pneumonia fatalities are high, there is a long history of inadequate health care resources available to the community. The idea that these researchers had developed was to arm community workers with electronic stethoscopes that would record the sound and send it to Johns Hopkins physicians, who could determine whether pneumonia was present and prescribe appropriate treatment. While a great idea, the implementation was proving problematic. The stethoscopes were picking up the background noise, for instance, the child crying or a truck driving by, and the quality of the sound being received at Johns Hopkins just didn't permit a diagnosis. This was happening well over half the time, rendering the procedure infeasible.

When the talk ended, I approached Jim Tielsch and introduced myself. In my typical confident manner, I said that I was sure we could solve their noise problem. That was the beginning of my third involvement with stethoscopes. In concept, noise control for a stethoscope is not unlike noise control in headphones and this is a problem that has been solved for at least a decade. Noise cancelling headphones have at least one microphone that records the external sound and then uses an algorithm to remove that noise from the noise+signal stream. In its simplest form, sound pressure exactly inverted from the noise is added to the sound in the headphones, thus cancelling out the noise. We needed to do something similar, to cancel the noise picked up by the stethoscope.

Of course, the stethoscope problem is far more complicated than the typical noise cancelling headphone problem, because the noise in the stethoscope case tends to be nonstationary. It could be a truck motoring by, a baby crying, or speech in the background. All these sounds change significantly in time, making their elimination much more difficult than, say, HVAC noise or the noise in an airplane, which tends to be fairly constant in intensity and frequency content and sound source location.

After my conversation with Jim Tielsch, I was introduced to Dr. Eric McCollum, a pediatric pulmonologist and Hopkins doctor active in the countries of Bangladesh and South Africa. Eric showed up with 25 or so electronic stethoscopes the team had been using and asked what we could do to help him. I took this project to faculty colleague Mounya Elhilali, and together we found a couple of interested students to start working on the problem. Leading the effort was Dimitra Emmanouilidou, a graduate student working under Mounya's supervision, and Ian McLane, an undergraduate in the ECE Department who was looking for a project to do with me. In our first tests, Ian mounted a microphone on top of each electronic stethoscope part that contacts the body (the bell). The newly enhanced stethoscopes were returned to Eric, who took them to the field.

As Eric gathered signals simultaneously from the stethoscopes and from the microphones mounted on them, the pairs of signals were given to Dimitra, who worked on algorithms to significantly reduce the noise from the stethoscope signal using the electret microphone to determine the background noise. The short answer to the question posed was yes, we could clean up the signals significantly. Our ragtag experiment suffered from some serious problems, notably repeatability, but we had evidence that we could clean the signals from the stethoscope. Because we taped the electret microphones onto the stethoscopes, and they might vary some in exact positioning and even move during use, we were limited in the precision we could achieve. Nonetheless, we had shown that by building a new stethoscope front end, the problem of noisy signals could be resolved.

While it was great that we had determined we could clean signals, it meant that a new device was needed. Designing such a unit went beyond fundamental research into product design and would require establishing a company to do the work. We did eventually do this, establishing Sonavi Labs in 2017. The company would focus on designing a new stethoscope that could be sold commercially and would be responsible for clinical trials. Sonavi Labs made use of our research and occasionally the line between our laboratory research and the company was hard to find.

As is usually the case in research, the answer to the first question simply led to other questions. Two in particular came to mind immediately. First, could we figure out what the doctors were focusing on when they listened to the chest sounds and decided that it did or did not sound like pneumonia? If we could, maybe it would be possible to train the stethoscope to provide a preliminary diagnosis regarding pneumonia and to get medication started earlier for sick children. This could be important as the cycle time in the process of going to JHU docs and then back to a remote location was often days and that raised the specter of pneumonia deaths that could have been prevented with earlier treatment. Second, the noise reduction algorithm was certainly computationally intensive. Might there be an alternative way to eliminate background noise, possibly with a stethoscope head that better matched the characteristics of the human body and naturally tended to suppress airborne noise? Ultimately, we worked to answer both these questions.

To answer the question of what doctors were focusing on for a diagnosis of pneumonia from a lung sound, we needed to get very clean signals. We did this, ultimately, using a few steps of processing. First, we processed the sound to suppress the background noise, additionally removing any sound from the patient crying or from friction due to the stethoscope bell moving on the skin. We also removed the heartbeat by using wavelet analysis in which we supposed there were two different sound signals—heartbeat and breathing—and found the most likely separation of them into two signals. We could then remove the heartbeat by subtracting that signal, leaving us with a clean lung sound signal.

Next, we put the lung sound through a filter bank meant to mimic the human ear, so that the signal would sound as heard by a human. From this we could generate an auditory spectrogram, a graph of the sound intensity at each frequency band (color coded) at each instant in time, displayed as a function of time—in other words snapshots of the sound frequency content as a function of time. Finally, we needed to process the sound through the equivalent of the auditory cortex of the brain, the place where humans process sound. We did this using a model of the auditory cortex that Mounya's team had built over the years. The final output, then, would be as close as we could get it to the stethoscope lung sound as perceived by a human. Armed with these well-processed sounds we could now ask what it was that was producing diagnoses of pneumonia by doctors. Ian McLane took on this question for his PhD dissertation work.

Through my time at Bell Labs, I was familiar with the issue of pattern recognition because a large fraction of the Acoustics Research Department I was part of worked on speech recognition. In the early days of speech recognition, the idea was to identify characteristics that defined various sounds and look for those characteristics. Typically, an algorithm would identify what sound a signal seemed closest to matching based on the model of these specified sound characteristics.

Today, instead of classical pattern recognition approaches, we use artificial intelligence to establish an algorithm. In the artificial intelligence approach, we don't start with preconceived ideas about what characteristics will define the best model. Instead, we gather lots of data—acoustic signals and expert judgements about them—and let machine learning tools develop the model we can use to determine a judgement on a new signal. The advantage of the machine learning approach is that it can find a model for recognizing a pattern that we might otherwise never have found. The disadvantages are that it requires training with lots of carefully curated data and even after the model is established, we might not be able to connect judgements to easily stated physical characteristics.

Ian McLane took the acoustic lung sounds that Eric McCollum was gathering in Bangladesh as well as his diagnoses regarding pneumonia and put them through a variety of machine learning algorithms. He used most of the data to develop a model, and then looked at how it did at predicting a pneumonia diagnosis on the remaining patients. For this set of data, Ian was able to predict a correct diagnosis of pneumonia in the data not used for training 91% of the time using the model he ultimately developed. While he wasn't able to give a name to what the characteristics were that Eric McCollum was listening to, it was clear that it was possible to use machine learning to identify likely pneumonia in children in Bangladesh. Ian lives in California and has established a company working in the machine learning general area.

The second question, whether we could reduce noise simply by changing the lung sound sensor to better match the characteristics of the body, was taken on primarily by Valerie Rennoll, a new graduate student in my team. The idea we had here was based on our knowledge of sound wave propagation at interfaces between media. If sound strikes an interface between two media that have very different acoustical impedances, then almost all the sound is reflected at the interface. Acoustical impedance is the product of the density of the material and the speed at which sound travels in that medium. Air and water, for instance, have very different acoustical impedances which is why it is so difficult to hear people speaking near a pool if you are underwater. Nearly all the speech reflects at the surface of the pool.

I was familiar with the Navy using impedance matching on their underwater transducers. Piezoelectric hydrophones often use lead zirconate titanate as the sensing material. It is a ceramic material much harder than water, so it tends to be inefficient at sensing sound as most of the sound hitting it reflects. The Navy solved this problem by wrapping some sensors in layers with impedance progressively approaching that of water.

The idea we pursued was to match the lung sound sensor to the impedance of a human body as closely as possible so that it would be much better at picking up sound from the body than from air. The motivation for doing this was clear. The signal processing developed by Dimitra to remove noise from the lung sounds was computationally intensive, making it potentially very expensive to incorporate into any product. If impedance matching to the body would mean selectively receiving sound from the body rather than the air, might it be possible to eliminate this hefty signal processing to clean signals?

Valerie was able to build a prototype sound sensor which was very similar in acoustical impedance to a human body. She showed that the net result was that noise suppression from sources in the air surrounding that human was not required for production of a signal clean enough to use for machine learning of signal patterns. This work was patented, and her device has led to some interesting offshoots. For instance, a student at Peabody Institute, the music school of JHU, has used her sensor approach on a bass violin to capture the full spectrum of sound it creates more effectively than a microphone in air near the instrument. Additionally, Hee-yun Suh earned her master's degree working with me to use Valerie's device as a throat sensor. It was mounted on the outside of a person's throat and could pick up the speech quite well. This sensor would miss some of the sibilant sounds (repeated hissing or hushing sounds like "sssss") but otherwise did very well. One could imagine a throat sensor being of use in very noisy environments such as might be found in some heavy industries and on battlefields.

I also have had a PhD student who is building on Valerie's work. Helena Hahn worked on building arrays of sensors each of which is matched to the body's impedance. Arrays permit us to be directional, so the aim was to be able to detect sounds from specific locations in the body. Helena is likely the last of my PhD students.

Upon graduation, Valerie joined APL. Hee-yun Suh moved to Princeton University to earn her PhD.

When the COVID-19 pandemic struck, we had an opportunity to expand our disease repertoire beyond pneumonia. As President Trump came down with COVID-19, there were suggestions that you could tell he had the virus from his voice. Drew Grant, a student who came to work with me, pursued research on whether, indeed, one could determine COVID-19 diagnosis based solely on auditory signals. Drew was able to find databases of breath sounds, coughs, and even speech of people with and without COVID. He found, using machine learning, that you can detect COVID-19 from speech, from coughs, and from breath sounds! Further, the computation involved in accomplishing this was sufficiently small that it could be put on a cell phone and produce nearly immediate results. We tried to get this patented as one could see that just having someone speak would permit screening of people showing up for a large audience event. Unfortunately, the process for filing patents is so slow at Hopkins that we were stymied and others presented similar technology ahead of us. I have not seen the technology developed into a simple screening tool that is sold commercially.

Drew is an especially interesting former student for what he represented. As my second Black student, he was a graduate of Morgan State University, where his advisor, Kevin Kornegay, was someone I had mentored in the CRFP program. Thus, Drew is an example of the strong Black community continuing to grow upon what the CRFP established. Upon completing his PhD, Drew went to work at Northrop-Grumman. He now works at APL.

At the time when many of my students were working on lung sounds, I had two students working in areas unrelated to that subject. Adebayo Eisape (Bayo), another Black student, started in my lab as a sophomore, and I couldn't get rid of him (nor would I have wanted to)! Bayo is very good with his hands and mechanically inclined. He worked on a problem that has long interested me—the issue of energy harvesting from polarized polymers. We've known for decades now how to polarize materials and how to use them as sensors by suspending them over a fixed metal backplate. In the sensor mode, we are looking for small signals that correspond to the vibration of the sensing membrane. In theory, one ought to be able to extract energy from the vibratory motion but attempts to do so have been largely unsuccessful. Bayo was able to create a membrane with a monopole (single sign positive or negative) charge distribution. When he suspended that membrane between two electrodes, the motion of the membrane from sound impinging on it would cause a voltage. Bayo was able to gather the small current from this motion and store energy in a bank of tiny capacitors. We started with low frequency energy as there is so much sound there due largely to HVAC systems. Bayo was able ultimately to gather milliwatts of power using this approach. Upon graduation, Bayo joined APL.

I also took on a student from Peabody Institute, Yoshimi Hasegawa. I asked Yoshimi to monitor the noise in the new PICU of Johns Hopkins Hospital and compare it to what we found earlier at the old PICU. Interestingly, the traditional sound metrics were nearly unchanged, but the perceptions of staff were that the new PICU was overwhelmingly quieter. This led credence to the work of colleague Erica Ryherd, who has shown that the occurrence rate of sound is generally more indicative of perceptual reactions in the hospital. The occurrence rate is a measure of what fraction of the time the sound exceeds a given level. Two units with similar traditional acoustic measures might have very different peak levels, or levels exceeded 1–5% of the time. The unit with more peak levels will always be judged as noisier. After finishing her MS degree with me, Yoshimi joined the University of Nebraska-Lincoln to earn a PhD with Erica. There she worked on a clustering analysis of the sound in pediatric units. She then moved to Australia to work. Several of the students are shown with me at my home in Fig. 18.1.

Although my publishing has slowed down some, and I am less able to travel so I am less seen at professional conferences, I continued to receive recognition for my work in the 2010s.

In 2012, I received the Icon Award of the Associated Black Charities, a 40-year-old organization dedicated to eliminating the barriers caused by structural racism in Baltimore. This award pleased me largely because it showed that I had become entrenched into my relatively new home of Baltimore and recognized for my equity work. That same year I was also asked to be the keynote speaker for the Program for Acceleration in Careers in Engineering (PACE). PACE seeks to prepare Black and Latino high school students for careers in the STEM fields.

In 2013, I received the Lifetime Service Award of the National Society of Black Engineers (NSBE). Also in 2013, I was made a Fellow of the National

Fig. 18.1 Some of my students joining me at home

Academy of Inventors (NAI). NAI is a professional society created in 2010 to provide support for people who are inventors and to encourage invention in the US. It now has over 4600 members, of which 2068 are Fellows, the highest grade of membership. Like the National Inventors Hall of Fame, NAI partners with the US Patent and Trademark Office.

In 2014, I received the Joseph Tyler Award of the National Technical Association. I was also presented with honorary doctorates from Princeton University and Temple University.

The decade of the 2010s brought me wonderful challenges and fabulous students to seek solutions. It was forged through a most satisfying connection between JHU medical personnel and engineering school faculty. I believe the work will be a significant part of my legacy.

19

Sonavi Labs

As Mounya's and my students made progress on the problem of pneumonia detection from lung sounds alone, it became clear to me that the world needed a new device for this purpose. Although I'd turned down the opportunity to commercialize inventions in the past, this time it seemed too important to pass up. We weren't just talking about better quality listening or less noise. We were talking about saving lives. I felt obligated to pursue device design and commercialization so that children living in resource-poor regions of the globe might be spared. The problem these thoughts presented was not unlike the issues I'd confronted before. I am not particularly suited to industrial design. There were still lots of unanswered questions about the science and I wanted to pursue them while others worked on the design of a new device to take advantage of what we were learning.

I discussed this with my collaborators, and they agreed that we had a technology worth pursuing. Mounya, Dimitra, Ian, and I formed a company. I served as President, Mounya as Chief Financial Officer (CFO), Dimitra as Secretary, and Ian as Chief Technology Officer (CTO). Ian suggested the name Sonavi Labs. We needed to find a CEO who could be full time working for the company. Without it, it was clear the startup would not succeed. I leaned on Ellington to consider the opportunity, and talked with her about the opportunity Gerhard and I had forgone at Bell Labs when we invented electret microphones. Ellington agreed to quit her sales job for a regional spine clinic and to assume the role of CEO. She is shown in that role in Fig. 19.1. The company was officially formed and registered in Delaware in October 2017. Because the company was based on work done at JHU, the

J. E. West, I. J. Busch-Vishniac, *Mic Drop*, https://doi.org/10.1007/978-3-032-20922-1_19

Fig. 19.1 CEO of Sonavi Labs, Inc. Ellington West

university was awarded a 10% ownership of the company, with the remainder being divided between the founders.

The creation of Sonavi Labs posed a seemingly endless number of problems. Among the thorniest turned out to be conflict-of-interest issues. These took two forms. First, there was the problem that Mounya and I could not simultaneously be supported on research related to the stethoscope and have a vested interest in Sonavi Labs. Federal rules strictly prohibit this to maintain research integrity. It's easiest to see the need for this in the pharmaceutical arena. The government wouldn't want to have drug efficacy tests put in the hands of people who own a part of the company producing the medicine as the temptation to misinterpret or fabricate data would be too strong. To solve the problem, Mounya and I both withdrew from the company in entirety after several months of helping set it up. The company shares awarded to Mounya were split between Dimitra and Ian. My shares were given to Ellington. Mounya and I continued as science advisors without pay.

The second conflict-of-interest issue involved students being engaged in the company workings. Dimitra's PhD work was somewhat distinguished from the work to create a device, so it was reasonable to argue that there was no conflict of interest with her role of Secretary. Ian, by contrast, was at an earlier stage in his PhD pursuit and his desire to serve as the Chief Technical Officer (CTO) of Sonavi Labs was routinely blurring the lines between his research and the company. Ian pleaded his case with university leadership and continued in both roles, although the university was unhappy with his dual role. The issue of Ian's conflict of interest was brought up multiple times prior to him completing his PhD. It even emerged as an issue toward the end of his PhD work, when his committee demanded to know exactly what he had done for a PhD and what he had done for the company. That Ian had collaborated with Valerie and Drew on their work did not help resolve this issue.

At the point that Mounya and I retreated from positions with Sonavi Labs, it became clear that setting up a startup company would require more than Ellington and the remaining two students could possibly provide. We needed a COO (Chief Operating Officer) and a CFO (Chief Financial Officer). Ellington brought on Arthur T. Ward, Jr. as COO and Jaishree McLane as CFO. Arthur had dated Ellington off and on and came from a well-connected and well-off family in Baltimore so he had significant connections to the funds we would need to raise. Jaishree (Jai) is Ian's mother. She had started and run a company in the health services industry in California. On the one hand, it made sense to pursue these two people for the positions. They knew us and we could persuade them to help without promising more than we should. On the other hand, as time would show, having close relationships between the company officers created all sorts of problems associated with bias and nepotism. It drew the attention of potential investors in a negative manner.

With a team assembled, Sonavi Labs managed to compete for a spot in a Baltimore City technology incubator. The team began the process of setting up a business plan and creating a new electronic stethoscope. From the start, the company saw itself as a socially conscious company, aiming to make health monitoring and diagnostic technology accessible. The hope at that point was for a product that would cost customers between \$100 and \$150. A decision was made to negotiate a contract with Harbor Design and Manufacturing for engineering and regulatory help, and to pursue NIH funding to support the research and development needed to produce a product.

By the end of 2018, the team had grown to include Ilene Busch-Vishniac as the Chief Innovation Officer (CInO), Brandon Dottin-Haley as Chief Business Development Officer (CBDO), and Dan McLane (Ian's father) through his company PhaseMargin to focus on product design. Dimitra had graduated and taken a job with Microsoft Research working in a competitive area, so she resigned from Sonavi Labs. Work on product design was ongoing with a planned launch in 2019 of the first product, a smart stethoscope. Fundraising, marketing, and regulatory strategies were being firmed up and a business plan developed. Arthur Ward resigned in 2019, after a series of arguments with the McLanes.

The business case for a smart stethoscope focused on respiratory diseases was not difficult to make. Besides pneumonia, there are three other respiratory diseases that too often lead to death or chronic illness. Asthma is the most common chronic disease of childhood. Chronic Obstructive Pulmonary Disease (COPD) is the third leading cause of death worldwide. Tuberculosis is the world's leading cause of death by infectious diseases, far outstripping HIV and AIDs. Taken together, there are roughly 800 million people worldwide searching for a more efficient and effective way to identify and manage their respiratory conditions.

The estimated cost per patient per year for management and treatment of respiratory diseases (in the US) stands at $6246 for COPD, $3266 for asthma, $235 for pneumonia, $34,600 for TB, and $15,571 for cystic fibrosis in 2021 dollars. The annual cost for the diagnosis and management of respiratory diseases worldwide is roughly $3.5 trillion dollars. In the US, nearly $20 billion is spent on the treatment of pneumonia and influenza, $56 billion on asthma patients, and $72 billion for treatment of COPD for a total of $148 billion from these three ailments alone. The greatest drivers of this cost are the current expensive methods for diagnosis and disease management and the high readmission/recurrence rates.

Additionally, while there are objective measures for many medical conditions, such as blood pressure, or glucose level, there are no objective measures that accurately indicate respiratory diseases. Even chest x-rays, the gold standard for diagnosing pneumonia, are read by doctors and thus subjective. Estimates are that doctors presented with a chest x-ray only agree that pneumonia is or is not present about 75% of the time. I admit I felt much better about medicine before I learned that statistic.

Compared to commercially available digital stethoscopes the Sonavi Labs devices would have a few major advantages. The device would provide onboard automated detection of abnormal lung sounds in real time, making it possible to aid in diagnosis. We learned quickly that the FDA and the medical community would not easily permit us to say that a device was capable of diagnosing, so we were careful to explain that our aim was to aid in diagnosis. Nonetheless, our goal was for our stethoscope to identify abnormal lung sounds and provide a probability that they indicated a particular medical issue such as pneumonia or asthma.

The new stethoscope would use adaptive noise suppression producing high-fidelity sound in any environment, including the noisy clinics typically found in under-resourced countries. In addition to removing background noise, the stethoscope would be able to remove the sound of a crying patient and the noise of coughing and of the device rubbing on the body.

The stethoscope would use multiple microphones to lower the sensitivity to body placement thus reducing the need for placement training. It accomplished this by using five microphones below the bell diaphragm to sense the sound rather than one.

Finally, the stethoscope would accomplish all of this with on-board processing so additional equipment would not be required. Mobile apps and a cloud platform would allow for personalization of treatment and disease management programs and integration with electronic health records. The company believed its respiratory assessment device would improve clinical workflow efficiency and decrease healthcare costs, provide greater access to healthcare for chronic disease patients, and decrease emergency room and urgent care visits.

Because work on the device matching the body impedance was in its infancy when Sonavi Labs was formed, the decision was made to pursue the first product line using a stethoscope bell that incorporated noise reduction and that resembled in form the current crop of stethoscopes. The idea was to introduce a body impedance-matched product line subsequently. With perfect hindsight vision, this was likely a mistake, as the costs in engineering, product design, and components were too high for the noise reduction.

The Sonavi Labs idea was to produce its first product, which it called Feelix, in two forms—one for doctors that was essentially a smart stethoscope, and one for patients to use at home that would only have the stethoscope bell and not the tubes to the ears. The team argued off and on about whether the stethoscope version needed to look like the old analog stethoscope, with rubber tubes connecting the bell to the medical professional's ears. Some of the team argued that the medical profession is very slow to change and that it would be hard to sell an electronic stethoscope they couldn't drape around their neck. After all, the stethoscope is an instrument that is more than 100 years old and has changed little in that time. That said, some of the competing electronic stethoscopes had the old form, while others chose to just provide the sensor and either a speaker or a Bluetooth connection to earphones.

This stethoscope form argument was one that I could contribute to, as I knew that the tubes distorted the sound. It was my suggestion that the company do away with the tubes and the distortion. The objection was raised that doctors and nurses train by listening to the distorted sound so we would be asking these folks to retrain. I suggested that we measure the distortion by monitoring the sound before and after the tubes are put on and then introduce that distortion electronically into the signal with an ability to switch it on or off. If on, the stethoscope would sound like an old-style stethoscope. If off, it would be lacking the distortion. Ultimately, my students did measure the distortion, but Feelix was not built to have the switch I suggested.

The strategic plan for the company changed significantly over a couple of years. At one stage it considered focusing on global health by outfitting community workers in low-resource nations with Feelix and gaining funding through nonprofit organizations and government agencies. Later it focused on sales to physicians and nurses, and then later still home health care and telehealth applications supported by the Center for Medicare and Medicaid Systems.

The strategic plan for the company that eventually emerged had multiple sales paths. First, stethoscopes would be sold to pulmonologists as a replacement for their current stethoscope. Additionally, pulmonologists would prescribe the home version for their patients for remote monitoring through the platform. This would work on a subscription basis, much like oxygen provided for those with chronic lung disease and would allow doctors to bill for remote patient monitoring. Patients would regularly gather lung sounds which would be screened by their doctors more frequently than a patient would normally visit a clinic. The platform would automatically signal doctors when abnormal lung sounds were detected, making early diagnosis of a problem faster and simpler. The aim was to focus on Medicare and Medicaid patients first. Obviously, this strategy meant that Sonavi Labs had shifted from a focus on under-resourced nations, to a focus on the US.

A second stream of funding would come from routine use of Feelix in telehealth visits so that doctors could hear lung and heart sounds much as they would in a standard patient visit. Our focus on lung sounds had forced us to develop an efficient means of separating out heart sounds from the lung sounds. In doing so, we had determined how to identify heart sounds and could easily play them as well. The main challenge with heart sounds is that they are not necessarily strictly periodic. Heartbeats speed up and slow down, so the technology needed to be able to adapt to these very human variations. Again, because the codes were created for doctors to charge for remote monitoring during the COVID-19 epidemic, they could prescribe Feelix and charge more for the telehealth visit. Brandon had negotiated a partnership with a telehealth provider with more than five million patients to pilot such a program.

A third revenue stream would come from specific disease targets such as in the detection of pneumonia or tuberculosis. Pneumonia is a major source of mortality among the very young and the very old. Placing Feelix in urgent care clinics would permit rapid indication of potential pneumonia infections without requiring an x-ray. Sonavi Labs imagined partnering with the wellness centers growing in drug stores so that patients could get a rapid pneumonia test.

Tuberculosis testing typically requires access to high quality wet labs and these can be in short supply in the low-resource regions where the disease tends to thrive. Replacing this need with an acoustic diagnosis would dramatically lower the cost of diagnosing tuberculosis by at least an order of magnitude in low-resource countries and aid significantly in the World Health Organization's goal of eliminating tuberculosis.

COPD patients would, using Feelix, be able to determine whether they were trending toward a serious event and modify or get medical treatment to avoid it. Similarly, cystic fibrosis patients would be able to determine whether they were suffering from mucus buildup requiring new treatment.

Finally, with these successes, Sonavi Labs was sure that insurers other than Medicare and Medicaid would be willing to include prescriptions for Feelix. The company calculated that this strategy would lead to revenues in excess of $100 million after only a few years.

At the time the strategic plans were developed, Sonavi Labs had some competition, but no one had products close to what Sonavi Labs was proposing. The market for stethoscopes then and now was dominated by Littmann, a 3M company, that had been making the stethoscope used worldwide for decades. ThinkLabs had developed an electronic stethoscope, but it did not have noise suppression or machine learning algorithms to aid with diagnosis. Eko created a line of electronic stethoscopes including clinical analytics, but they did not have noise suppression incorporated, so their analytics were suspect as they were dealing with noisy signals.

From 2018–2020, Sonavi Labs made good progress in some areas, such as funding and clinical development and struggled in other areas such as product design and FDA approval. In 2018, Sonavi Labs left the ETC technology incubator space and moved rent free into space in Harbor Designs.

Ian and Ilene wrote two NIH STTR grant proposals, partnering Sonavi Labs with Johns Hopkins University (Mounya and me). The first was aimed at producing Feelix and proving its value on diagnosing pneumonia. This required Feelix to detect crackles, which tend to show acoustic energy in the

100–500 Hz range. (For context, a symphony tunes to 440 Hz). The second grant focused on expanding the diagnostic capabilities of Feelix to asthma and COPD exacerbations. For asthma and COPD, Feelix would need to detect wheezes, which are typically between 100 and 2500 Hz. Both the grant proposals were funded—a hit rate that is rare in the biomedical research field.

Working with Dr. Eric McCollum, Sonavi Labs was able to show that it could diagnose pneumonia from lung sounds with better than 90% accuracy for a set of Bangladesh sound files used both for training and testing, although it did not show that this model worked on pneumonia sounds gathered from other locations. Ironically, the COVID-19 pandemic made it nearly impossible to do any research at US hospitals, as the staff were overwhelmed with treating patients, so testing in the US was delayed. This was unfortunate as we were certain we could diagnose COVID-19 quickly given the chance to incorporate machine learning of the lung sounds.

An additional project that Ian worked on with a hospital in Belgium focused on the use of Feelix technology for cystic fibrosis patients and proved very useful in showing the impact of percussive mucus clearing treatments. In this research, patient lung sounds were measured before and after a treatment involving thumping on the chest to loosen mucus. Results were able to show significant changes in the lung sounds accompanying the changes in trapped mucus as determined by doctors. This study went a step beyond all others and was able to determine the characteristics of the sound that the machine learning had identified as indicative of an issue. It was primarily the ratio of the sound energy in the high frequency bands compared to that in the low frequency bands that was indicative of mucus buildup. This determination is important because it eliminates the need for a complicated machine learning model and allows for the direct measurement of the sound energy ratio instead. This cystic fibrosis work also developed a means of determining the respiratory rate from lung sounds and reported a 94% accuracy compared to the assessment by physicians.

The bottom line is that the fundamental basis for the Sonavi Labs products was being well established and reflected a very important upgrade to what was on the market. In other words, the technology was primed for success.

Ellington was terrific at pitch presentations and regularly won technology startup competitions, including one in Dubai at the Arab Health conference, the world's largest medical exposition. She also won the pitch competition at South by Southwest (SXSW) and a host of other venues. I was and remain so proud of her for developing this skill and succeeding so often. Ellington's pitches made it much easier to raise funds, which Sonavi Labs did in five separate convertible note offerings for a total of about $6.5 million. While the first of these offerings was mainly friends and family of Sonavi Labs team members, later offerings included some technology funds. At one point, Sonavi Labs even had an offer to be acquired, but the offer seemed far too low to be accepted and the company that tried to acquire it went bankrupt themselves just a few years later.

Sonavi Labs filed for patent protection for their designs and worked with an excellent law firm to push these patents through. Outside of the patents that Johns Hopkins had pursued and that Sonavi Labs licensed exclusively, there are eight additional patents that the company has been granted. These patents are what prevent competition from abroad from entering the US market with devices similar to Feelix.

At one of the pitch competitions, Ellington made the acquaintance of Leslie Wise, a serial entrepreneur in the medical space. Leslie became an advisor to Ellington and officially joined Sonavi Labs as a board member. Through Leslie, Sonavi Labs developed the strategic plan of targeting Feelix for home use through prescriptions of doctors and applied to have the device covered by Medicare and Medicaid. There was only one somewhat similar device on the market that had such clearance so this would be a big competitive advantage. The first attempt to get clearance for the coverage was not successful, but that was largely due to the cost of Feelix (at the time estimated to be about $400). The company remained certain that this was the right strategy and that a lower price point would ultimately let Feelix pass muster.

All medical devices require FDA clearance and Sonavi Labs struggled to get this clearance. It was early days in the artificial intelligence revolution and the FDA examiners were generally not inclined to approve new products which relied on machine learning. Feelix was ultimately approved for producing clean lung sounds but not approved for suggesting diagnoses to physicians. Two attempts to get such approval failed.

Although Sonavi Labs raised a significant amount of money, it was insufficient to support the incredible legal costs that come with producing medical equipment. Lawyers had to be hired to establish the business, to handle funding solicitations, to produce patent applications, to produce the book volumes that passed for FDA applications, and to deal with potential international partnerships and contracts. In the end, legal costs were siphoning far too much of the funds at Sonavi Labs's disposal.

Despite proximity to Harbor Design and Manufacturing and the close, personal relationship with the engineer designing and assembling Feelix, we never managed to get devices that were working reliably. There were problems with noise leakage within the device which were partly solved by blocking the power port, and the connection to the software to save and manage recordings, created by Jai based on prior software she had sold, was simply unreliable. The episodes of finger pointing and name calling escalated as frustrations grew and trust between teammates was eroded. The relationship between the McLanes and Harbor Designs became strained to the point of breaking.

By 2021, it was clear that Sonavi Labs was not making appropriate progress on producing a product for sale. Complicating matters, Dan McLane died suddenly of a heart attack, leaving the company with a set of partially working devices but no clear notes on how to make more or guidance on function of the electronics. At this point things started to look dire. Funding was beginning to become scarce, the product was not working reliably, and transparency had become an issue. Although the company muddled along for a bit, things came to a head toward the end of 2022.

Starting in the middle of 2022, Ellington hired Eric Solander to redo the software supporting Feelix's recording and management of audio signals. Eric reported that the software currently in use was not HIPAA compliant, i.e. not legally compliant with privacy laws. This caused an enormous pushback from Ian and Jai, but they ultimately admitted that there were issues that needed to be addressed. Further, although Ilene was the principal investigator on both NIH grants, Ian was not inviting her to research meetings and Jai was not making financial transactions transparent to her. Worried about her liability, Ilene issued a demand that she be invited to all research meetings and that any purchases on grant money require her prior approval. Jai and Ian said they would resign if Ilene's demands were met, and Ellington decided that losing Ian and Jai was more detrimental to the company than losing Ilene. Within 24 h, Ilene had quit Sonavi Labs and told NIH that she could no longer serve as the principal investigator of the NIH grants. Unfortunately, there was nobody on the Sonavi Labs team that could take her place, so the remaining funds on the grants were lost to the company.

Just a couple months after Ilene left the company, Ian and Jai resigned as well. That left the company with no technical expertise, very little funding, and a set of convertible notes about to be due. Fortunately, all but two of the investors agreed to take stock in the company rather than to be repaid their investment with the promised interest. It is an understatement to refer to the situation as a big mess.

There are many reasons why startup companies fail. Sometimes, the technology is simply inadequate, but that was not the case for Sonavi Labs. Sometimes competition catches up too quickly, but again this was not the problem for Sonavi Labs. Most of the time, a startup fails because there are personnel issues that impede progress. This was certainly the case with Sonavi Labs.

With 20:20 hindsight, I am comfortable saying that a few personnel mistakes were made in Sonavi Labs. Ian McLane is probably the brightest student I've ever had, but it was a mistake to allow him to simultaneously be a student and the CTO of Sonavi Labs. It presented too many conflicts of interest. Additionally, it was a mistake to recruit officers with personal relationships with the founders. We were required to reveal such relationships in our contractual documents, and it was an issue raised by some potential investors. It also occasionally made conflict resolution difficult.

After Ilene, Ian, and Jai left Sonavi Labs, Ellington reorganized the company. Dr. Eric McCollum began working more directly with Ellington rather than Ian and became a more active team member. Eric Solender brought in a team to entirely redesign the software for Feelix. Through Dr. McCollum, new grants were obtained which enabled progress on clinical validation. Money was scarce but the company shuffled along.

In 2024, Ellington made contact with a Polish company that created a device very similar to Feelix. Their device is called StethoMe and was created with active involvement of doctors and engineers. Unlike Feelix, it uses only the stethoscope bell, eliminating the tubes to the ears and using standard earbud connections. The strategic plan for StethoMe has been to sell it directly to patients to use at home, but recently the company has realized that they need more of an interaction with the practitioners diagnosing ailments. For this reason, they would like to pursue the approach Sonavi Labs has charted. The StethoMe device including the app is marketed for a price of between $50–100 but can't be sold in the US because it would violate the Sonavi Labs patents.

Talks between StethoMe and Sonavi Labs have led to a merger plan. The merged company is to be called Sonavi Labs, and the device sold will retain the StethoMe label. The merger allows for fundraising for the new, combined company to begin in 2025. The clever engineers and doctors who pioneered StethoMe remain part of the company. Original investors for both companies benefit from the merger.

I am very optimistic that the new Sonavi Labs will be able to thrive and do well. The Polish and American doctors working with the team will continue to develop machine learning models that will aid diagnosis and treatment of pneumonia, tuberculosis, cystic fibrosis, asthma, and COPD. The lower-priced device will likely gain traction with Medicare for home and telehealth use. That Sonavi Labs has managed to survive its near death and emerge as a company with a bright future is a testimony to Ellington's perseverance and talent. I am very proud of her. It is my hope to contribute to the future success of Sonavi Labs as my retirement from JHU is finalized.

20

Thoughts for the Future

One of the items that separates Black families from white families in the US is how we think about money. In my youth, the Horatio Alger story and the idea of pulling yourself up by your bootstraps was prominent, but it applied to white families and not Black families who were often kept from jobs that would permit them to rise in the financial hierarchy. For many white families, money is passed down from generation to generation. For most Black families, this was simply not possible when I was growing up. There generally wasn't a lot of extra money floating around in Black families, and the idea of saving was mostly a pipedream as banks discouraged Black people from establishing accounts. It made more sense to live for the moment and have something tangible like a home or a farm, that could be passed on.

I worked at Bell Labs for nearly 45 years and at Johns Hopkins for another 22 years. I was paid reasonably well for all those years, so I suppose I could have amassed quite a nest egg had that been important to me. In reality, I've never much cared about money except for making sure I had enough to do what I wanted. Over the years I have helped the kids and grandkids when they've needed it and put some of my money to good use for the community (as in helping establish and support the Plainfield Science Center). I've never gotten interested enough in wealth that I bothered to hire a financial advisor or to put money into the stock market. Now in my twilight years, I sometimes wonder where the money has gone. I am spending a good bit each month on caregivers who help me given my serious mobility limitations. I've few remaining funds in the bank. Should Sonavi Labs do well eventually, the funds from my investment in the company will be the inheritance I leave for my family.

J. E. West, I. J. Busch-Vishniac, *Mic Drop*, https://doi.org/10.1007/978-3-032-20922-1_20

I turned 94 years in 2025 and finally convinced JHU that it was time for me to retire. And while I've had plenty of time to consider my mortality, I confess it has rarely crossed my mind. Until I was nearly 90 years old, I didn't have a will. It just didn't occur to me that I should have one since I was still in good health. I understand that I won't live forever, but I intend to make the most of whatever time I have left, because there is still so much to do, so many unanswered research questions, so many students who I might encourage to consider careers in technical fields.

A few years ago, I approached people at Johns Hopkins to talk about retirement. They asked me to please not retire, so I made no change to my status. But now, I have students asking to work toward a graduate degree with me as their supervisor, and I am reluctant to take them on because I am not certain I can mentor them through completion of the degree. I have been given the title of Emeritus Professor and will continue to be involved in the department, but clearly, I cannot continue to carry the normal workload of a professor.

My grandmother instilled great confidence in her children and grandchildren. In general, I was taught nothing can stop me. If there's something that I wanted to do, I could get it done. I happen to be a reasonably good scientist, but I think that if I had decided to be a musician, I would have been among the best musicians as well. I believe that at the age when most people are willing to give up, I still have some fight left in me. There are still battles that I need to win.

I do have some projects that I continue to pursue. Notably, Drew, Bayo, and I are working on how to build a bridge from JHU's Whiting School of Engineering to APL. The problem seems to be that the overhead costs at APL are very high, so research funds obtained at APL don't deliver significant funds to Johns Hopkins and that seems to have blunted interest by top management in finding a path for greater interaction. We three are working on a means to encourage more interaction, as both APL and the engineering school would benefit from more collaboration. What we haven't yet defined is the project that we can use as an example of what a collaboration model might be. It will surely be sensors and actuators, but we need to be more specific than this before we try to get official sanctions for the project.

I am, quite naturally, thinking about my legacy. Of course there are the normal aspects to my legacy: four wonderful children, six grandchildren, and two great-grandchildren. Melanie has worked at the nexus of music and engineering, specifically using music and musical training to improve technology understanding. Her focus is on the significance of rhythm. She is married to a Greek man from Thessaloniki who is an electrical engineer, and they have two children. Laurie is a visual artist, a musician, and a data specialist for a marketing company, married to a man of Irish descent and they have two children. She is shown in Fig. 20.1 with Ian's daughter, Sana. Jay is an engineer at a helicopter company and is married to a woman whose family is from Barbados. Together they have two children. He continues to write and produce music and has a great potential to be a standup comedian. Ellington is married to an engineer of Irish descent. So, my four children have married spouses who they love, only one of whom is Black. While there are some who might take this to mean I failed as a parent, I take it the opposite way. I raised them with the freedom to love whoever they desired without restriction. And

Fig. 20.1 Laurie and Sana, Ian's daughter

while colorism still exists, the impact of being mixed race on my grandchildren has been far less than it was for me.

Among my grandchildren, Melanie's son Aaron is now an independent music producer. His sister, Kaliroë, has completed her PhD in physics from MIT and a postdoctoral fellowship at Columbia University. Laurie's kids, Ryan and Jessie, are doing well. Ryan, a creative, peaceful, loving soul taking a lot after Laurie, is living in California with his partner Angela and working at a dispensary. Ryan's daughter, Elowyn, and son, Zander, are my first great grandchildren! Jessie is living with her partner, Ricardo, in New Jersey and working for an insurance company. Like her mother, she is an incredible artist. Jay's kids are still young and finding their way. Both James and Julian were in and out of school. James has completed his associate's degree and enlisted in the Army reserve. As of 2025, he is working toward his BS at Rutgers University. Julian, a talented lyricist, is working at Atlantic Healthcare. Both James and Julian are involved in music and have home studios. Julian's speciality is rap music.

On top of my children and grandchildren, my legacy includes the usual markers of a research career: seven graduated students and too many patents, papers, and book chapters to count. It is the students I've mentored who I rely upon to carry science forward and to continue my determination to provide opportunities for people of all types to participate in technology innovation. As one of the most senior of these students, Dawnielle is clearly following in my footsteps by teaching students in materials science and by actively participating in programs to help women and people of color find technical career paths. And of course, the James West Postdoctoral Fellows of the ASA keep my story visible while opening the door for well-deserving students in my chosen field.

Clearly, my work on electret devices is the primary legacy of my research career. It resulted in commercial products that have been hugely successful. It is not an overstatement to say that this body of work changed the way we do things in acoustics. I also would like to be able to say that my work on acoustic detection of lung problems has saved lives, but we haven't managed to produce the needed devices and approvals yet although the technology is ripe for commercializing.

Beyond my contributions to science, I hope that my story sends a clear message challenging discrimination. Throughout my life I was shortchanged on opportunities and recognition simply because of my skin color. Nonetheless, I have been able to be a very successful scientist. Imagine what I would have been able to do had I been given access to the facilities and training others had open to them. I am certain I would have accomplished even more. I will even go so far as to state that with the advantages given to my white colleagues, I would have been a President of Bell Labs.

I think the most important thing for people to take away from my life history is that there's nothing extraordinary about me. I have lived a life not unusual for Black men of my age and origin, although I admit that my family has produced more academically gifted people than is typical. I hope that my story will serve as an inspiration for Black and brown people to pursue careers in STEM fields. They are not only rewarding, but also lots of fun, and obstacles that I encountered have been lowered if not completely eliminated. With more equitable training and education, imagine how students of color might perform now.

There are many times in my life when I was the "first Black man" to attend or to be something. What being the first did for me was to make me work harder on getting more underrepresented minorities and women involved. I don't consider myself that unique. I know plenty of women and underrepresented minorities that I consider much smarter than me, so if I am being honored, just imagine the number of overlooked Black men and women who should also have been honored.

I want to rid the world of discrimination, and I've worked hard on programs to contribute to this goal. I very much want young Black men and women to have opportunities that were denied me. This is especially true in educational options. Without education we can't produce great leaders, and the Black community has so much to contribute as leaders in every field.

We have come to a time in the US when diversity initiatives are being torn apart because they are said to undermine merit-driven actions. There is an assumption in this assertion that is totally wrong. Diversity initiatives never suggested that people should be hired, rewarded, or given opportunities purely based on their race, religion, gender, sexual preference, or physical ability. Instead, diversity initiatives aimed to provide the same opportunities for minority groups that white males had long had. The aim has been to provide the *opportunity* for an individual to demonstrate that they merit the job, promotion, or scholarship.

This view of diversity as a snatching of jobs and scholarships away from those deserving to hand them to people less qualified and awarded only because of their race, physical status, gender, or religion is absolute nonsense. I think of diversity, equity, and inclusion (DEI) as a continuum of what we are trying to achieve. Most activities benefit from having different perspectives expressed. To do that we need diversity. But having diverse perspectives does not ensure that people are treated the same regardless of their race, religion, gender, sexual preference, disability status, or religion. That requires equity—a fairness model in which people are treated equally and judged on their ability to perform. But even equity doesn't mean that activities will be crafted while thinking of how to reach everyone they might interest. That requires thinking about inclusion and what it offers to participants and how it improves performance.

Additionally, the argument about merit versus diversity initiatives suggests that one can determine merit objectively and that's far from true. We have never had objective measures of merit. We know, just as an example, that standardized entry tests for university are biased against people of color. We also know that there is essentially no correlation between these test scores and performance once students are admitted to a school. The point is that the merit argument is specious and itself designed to permit but obscure perpetuating bias. It is essential to fight against this nonsense with everything we have.

For me, it's all about change. What can I do to affect a change that is long-lived? There is a lot of distance between the segregation era I grew up in and where we are now, but we still have work to do to close the equity gap. My goal in programs I've worked on, some of which I created, has been to close those gaps in the STEM fields by establishing pathways for students who look like me to succeed.

The world has become an increasingly global and globally competitive place. For the US to retain a position of importance in the world, it must hold an edge economically and militarily. To do so requires the US to be a leader in science and technology. We currently have too small a segment of our population interested in pursuing technical careers and that poses a big problem. We need to make science, engineering, and math more accessible and attractive to a larger part of the population. And we know how to do that. Students need to see reasons to learn the science. Once inspired, there is no stopping students. What inspires them is the application of the science rather than the theory. It is this which has led to the development of a problem-solving based education approach rather than lectures. My early life would have been very different if problem-based learning had been popular when I was growing up.

We've allowed the idea that technical subjects are hard to learn to exist for far too long. In many cases, we've even prided ourselves on how hard it is to gain a degree in these areas or to be considered proficient. Some faculty still start their classes by telling students to look at the students to their left and their right, noting that in the end at least one of them will no longer be in the program by the end of term. This suggests that we are wasting resources on students who won't finish the program. Everyone has the right to change their mind and find their ideal pathway, but as educators our job should be to identify the obstacles in the road and either remove or lower them.

My experience is that parents have also contributed to this problem by telling their children that they didn't understand math in school and that's why they didn't pursue a career in math or science or engineering. It was just too hard and too complicated. More than once, I've told parents that this is not what they should be telling their children. It sends the message that it's okay for them to avoid math and science and label it as too hard.

We've historically excluded people of color and women from many areas of technology and continue to have barriers (cultural, economic, and social) to their full participation. This means we aren't taking advantage of all the talent that exists in the country, and we are doomed to be overtaken as a world leader if this continues. While we were among the leading producers of graduates in STEM fields through most of the twentieth century, we are now being eclipsed by a few Asian countries. In 2020, China produced 3.6 million STEM graduates, and India 2.6 million. That same year, the US produced 820,000 STEM graduates.

But it's more than just a numbers game. All of us bring our backgrounds with us when we work to solve problems—technical or social. By excluding women and people of color, we narrow the diversity of perspectives on any issue. The big technical problems of today won't be solved that way. We need as rich a set of perspectives as we can possibly find so that we consider all possibilities for success and identify all the ways we could fail. For that reason, we need a workforce that is diverse economically and socially. We need people of all religions, genders, physical abilities, and races to work together toward a common goal.

The last few years have produced an enormous backlash against diversity and equity initiatives. I am hopeful that we've come far enough away from segregation and slavery that retreat fully into the discrimination of the past is not possible. Nonetheless, the current politics of hating "wokeness" is demoralizing. It's interesting that the choice word of the political right in thinking about diversity issues is "woke." I grew up using that word and knew that it stood for "awareness of the issues facing Black people in the US". Friends would say we should stay woke.

As Black studies programs are being eliminated and efforts to enhance diversity and achievement curtailed, my friends and I are getting panicked phone calls from people many of whom have some connection to the Bell Labs CRFP program. They express their angst having seen that the demographics of the entering classes of MIT, Harvard, and other universities show far fewer Black students than just a year ago. They ask for our guidance in what they can do to right the ship. But my past experience has not been good in trying to help people solve their problems. It was one thing to work on diversity at Bell Labs or now at JHU and in Baltimore. It is another matter entirely to intervene from the outside of MIT or Harvard. In many cases, these phone calls are coming from people who didn't have the time or inclination to help when we were actively seeking assistance for programs we were growing. Now they want us to throw them a life raft. I'm afraid we don't have one to offer. They need to work the problem from inside, as we did. In other words, they need to accept responsibility for diversity and equity on their own.

I am often asked to give advice to those following me, and I tend to shy away from that because it is more meaningful for people to find their own way in life. That said, I do have some observations that I want to give voice to.

First, I note that no matter what your race, religion, sex, or color, working hard is the basis for advancement. There is no magic bullet. You must perform to be recognized and valued. Trying to be carried along in a group and taking advantage of something other than your merit, such as your race, just won't work. Eventually, your coworkers will object to the unfairness of the situation. Alternatively, as we are currently seeing, a political change can make your special characteristic, be it race, gender, religion, sexual preference, or physical disability, less respected as a reason for being given a special opportunity and you will be out of the job unless your value as a worker is seen.

Now, there is a flaw in this argument. Some people work hard and still aren't recognized and given credit for their efforts. Then, I would argue, you need to be aggressive. You must demand credit for your work and not allow others to deny it. I regularly had people telling me that they had thought of the devices or designs I was generating before I had. I told them that I'd gladly include them on the papers and patents if they would just provide me with the documentation. Not one person returned to claim their stake in my work after that conversation.

And this brings up another comment about how to deal with adversity. When I first got challenged by people telling me they had had the same ideas earlier and deserved credit, I pushed back forcefully and sometimes nastily. But I learned that this approach never brought the issue to a conclusion. It just produced repeated flare ups and hurt feelings. Over the years I learned that it is possible to be forceful and yet still be polite. By calmly requesting proof of prior work from people, I could respect them but forcefully define what it would take to make me give them credit.

Over the years of my career I learned that in science it is common for there to be a strict separation of professional life from personal life. It is possible for people to respect your scientific talents, without respecting you as a human being. This has been hard for me to fathom, since I am first and foremost, a human being and I assumed that respect for my scientific abilities would translate into appreciation of me as a person. Somehow, some people are able to find the nearly nonexistent boundaries between work and personal characteristics and to respond positively to one while responding negatively to the other. This means your work colleagues may not be your friends, and I think that is just fine. They don't need to be both. It is perfectly okay to have a working environment in which you are respected and treated fairly for your performance, even if you go your separate ways outside of working hours.

I have come to understand that if one looks at the record of accomplishment in science by Black people, it tends to focus on a few exceptional people. Further, the stories of these people often focus on their exceptionalism and how they overcame the odds. This sends the message that generally Black people have not been at the center of scientific thought and ability and that is not correct. What is a more accurate description of the situation is that while Black people have contributed as much as any group to important scientific advancements, their contributions have usually been overlooked or not credited to them. The exceptions simply could not be thrust aside. The point for students is that they don't have to be exceptional to have a wonderful career in science. They simply need to work hard and make sure they get credit for their work.

An example of what I mean when I talk about the contribution of Black people to science is in my area of telecommunications. There are now five or six Black inductees into the National Inventors Hall of Fame who focused on telecommunications and without their work, we either wouldn't have the pervasiveness, speed, and quality of communications we now have, or it would have taken significantly longer to develop. Besides my work on the microphone, there is Link Hawkins's work on a UV-resistant polymer to wrap the telephone lines, Marian Croak's invention of voice over internet protocol (VoIP), Victor Lawrence's work on modern digital signal processing, Kerrie Holley's invention of a software and programming architecture for large enterprises, and Mark Dean's work to establish a means of plugging in peripheral devices that enables the internet. Taken together, these inventors created a significant part of the scientific basis for modern communication. None of these were superhumans. All of them are or were ordinary people of talent who seized the opportunity to make a difference in flexing their mind. Imagine how many others we might have been able to list had the opportunities been presented equally to everyone with talent!

GPSR Compliance
The European Union's (EU) General Product Safety Regulation (GPSR) is a set of rules that requires consumer products to be safe and our obligations to ensure this.

If you have any concerns about our products, you can contact us on

ProductSafety@springernature.com

In case Publisher is established outside the EU, the EU authorized representative is:

Springer Nature Customer Service Center GmbH
Europaplatz 3
69115 Heidelberg, Germany

www.ingramcontent.com/pod-product-compliance
Lightning Source LLC
Chambersburg PA
CBHW060107210726
48503CB00021B/597

* 9 7 8 3 0 3 2 2 0 9 2 1 4 *